Contractor's Guide to the Plumbing Code

Contractor's Guide to the Plumbing Code

Publication Date: April 2001
First printing

ISBN 1-58001-072-5

Acquisitions Editor:	Mark Johnson
Cartoonist:	Jeremy Gregory
Cover Design:	Mary Bridges
Illustrator:	Mike Tamai
Interior Design:	Alberto Herrera
Manager of Development:	Suzane Nunes
Production Coordinator:	Cindy Rodriguez
Printed & Bound by:	Copy-Rite Press

COPYRIGHT © 2001
5360 Workman Mill Rd. Whittier, CA 90601-2298
www.icbo.org
The World's Leading Source in Code Publications

ALL RIGHTS RESERVED. This publication is a copyrighted work owned by the International Conference of Building Officials. All rights reserved, including the right of reproduction in whole or in part in any form. For information on permission to copy material exceeding fair use, please contact: ICBO Publications Department.

Information contained in this work has been obtained by the International Conference of Building Officials (ICBO) from sources believed to be reliable. Neither ICBO nor its authors shall be responsible for any errors, omissions, or damages arising out of this information. This work is published with the understanding that ICBO and its authors are supplying information but are not attempting to render engineering or other professional services. If such services are required, the assistance of an appropriate professional should be sought.

Comments on this publication are welcome and will be considered in future revisions. Send to products@icbo.org.

PRINTED IN THE U.S.A.

Preface

Contractor's Guide to the Plumbing Code is the first in a series of titles designed to familiarize contractors with current code requirements. It is written for the busy professional, whose goal is to provide a safe and affordable plumbing system that meets the minimum requirements of the latest plumbing codes. The "Contractor's Guide" is not a substitute for the code; it is designed to be used alongside the code, and to simplify many of its provisions by supplying explanations and illustrations.

Chapter 1 sets the stage by detailing an extensive background on current plumbing codes. Chapter 2 provides key definitions of terms in the current plumbing codes. Required documentation, permit issuance, and alternative materials and methods of construction are discussed in Chapter 3. Issues related to conventional drain, waste, and vent systems, such as sizing of the drainage system, venting, and installation of pipe and fittings, are covered in Chapter 4. In Chapter 5, water supply and distribution are discussed, as well as the important role of backflow devices. Chapters 6 and 7 conclude with discussions of the types and quality of plumbing fixtures, the installation requirements for these fixtures, and issues related to storm drainage.

An added feature of this guide is an appendix that includes a brief commentary, comparison, and cross reference between the new 2000 *International Plumbing Code*® (IPC) and the regionally based 1997 *Uniform Plumbing Code*™ (UPC).

The simplified approach to code provisions makes this guide an excellent reference for students and first-time code users.

Acknowledgments

All portions of the ***International Plumbing Code®*** (IPC) are reproduced from the 2000 edition, © 2000, and ***The International Plumbing Code: A Guide for Use and Adoption***, © 1998, with the permission of the publisher, the International Code Council, Inc. (ICC). Note that much of the material referenced in this document is also contained in the 1997 edition of the IPC because code changes between editions are rarely wholesale in nature.

All portions of the publication ***1997 IPC/1997 UPC: An Overview and Analysis,*** © 1998, are reproduced with permission from Professional Plumbing Seminars and the International Conference of Building Officials (ICBO).

The authors of this guide are David Cantrell, technical lead, plumbing/mechanical, County of Snohomish, Washington, and Robert G. Braun, II, building inspector, City of Woodinville, Washington, at the time of the writing of the book and formerly a private plumber with his own businesses in Washington and California. Both authors are licensed plumbers with decades of experience in both field construction and inspection. The staff at the Northwest Resource Center of ICBO in Bellevue, Washington, provided the initial preparation of this guide, including initial editing. The guide's illustrator is Jeremy Gregory.

Table of Contents

1 Plumbing Codes—Who Needs Them? 1
    Background 1
    Administrative Details 9

2 Definitions 13

3 Documentation, Standards, Alternatives and Inspections 29
    General 29
    Getting the Permit 30
    Documentation 30
    The Code Official 32
    Standards and Approvals 32
    Materials, Plans and Preliminary Layout 34
    Required Inspections 39

4 Conventional Drain, Waste and Vent Systems 41
    Typical Schematic 42
    Drainline Connections 43
    Sizing the Drainage System 45
    Cleanouts 47
    Venting 47
    Vent Termination 48
    Nonconventional Venting Methods 51
    Indirect Drains and Special Wastes 58
    Installing the Pipe and Fittings 59

5 Water Supply and Distribution 61
    Water Is a Good Thing 61
    Water Services 61
    Sizing Those Pipes 62
    Water Hammer 64
    Is Cheap Pipe the Best Way? 67
    The Installation 68

    Too Hot for Me . . . Hot Water 68
    Backflow Devices 68
    Healthcare Plumbing 70

6 Fixtures that Use Water 71
    Fixtures 71
    Number of Fixtures 72
    Other Fixtures 73
    Dishwashers and Showers 74
    Water Heaters 74

7 Storm Drainage 79
    General 79
    Materials 79
    Roof Drain and Overflow Design 80
    Specific Design Criteria and Details 81
    Scuppers and Leaders 82
    Combined Storm and Sewer Systems 83
    Below-grade Storm Drain Systems 83
    Engineer-designed Systems 84

8 Private Disposal Systems 85
    General 85

Appendix A—Comparison and Cross Reference: 1997 UPC and 2000 IPC 87

Appendix B—Which Is More Cost Effective, IPC or UPC? 135

Index 141

1

Plumbing Codes—Who Needs Them?

In this chapter, we will discuss the importance of plumbing, the history of how it has evolved over the years, the reasons for codes, the purposes of codes, how codes are produced, and the origin of the technical code provisions in the United States. The role of the building department in enforcing codes and reasons for permits are also discussed.

Background

Ah, the comforts of home, a nice beautiful home! You've been able to build your dream home and now you can enjoy the remainder of your days. Except that you've just discovered a couple of glitches. The toilet doesn't always flush and you keep calling the plumber to come out and snake the lines. And why do you get those terrible smells in your house?

These are just some of the problems that can occur if the plumbing is not installed correctly. That is why a well-working plumbing system is dependent on good design and the installation practices of those who have acquired the technical knowledge and skills to install it correctly.

But that doesn't necessarily mean that you have to be a plumber to accomplish this task. There are technical guides and other books that can help you and can teach you what to do. Modern, up-to-date plumbing codes are designed to provide minimum requirements for installation so we can have a lasting product that will bring us happiness and pleasure for many, many years.

Some History

Where did plumbing start? It may surprise you to discover that plumbing has been around in one form or another throughout much of recorded history. Written evidence has been traced as far back as Biblical times in Mesopotamian cities, particularly in Babylon, under King Hammurabi (see Figure 1-1)(O'Bannon 1989).

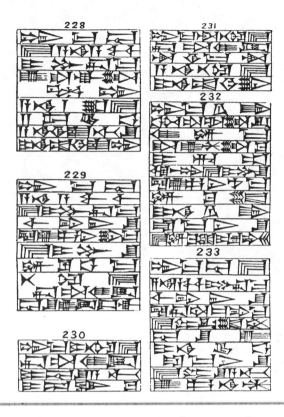

Figure 1-1 *The Code of Hammurabi*

In fact, some earthenware pipes, masonry sewers and toilets installed by the Mesopotamians in about the year 2500 BC are still in fair working order. The members of the ancient nation of Israel were instructed in their laws to go outside the camp along with a digging instrument in order to "take care of business," burying the evidence, so to speak. (Deuteronomy 23:13.)

This was obviously a very healthy practice. Disease would be less likely to spread in their community. This, along with other public health practices such as washing ones hands afterwards, has proven to be one of the most effective methods of containing the spread of disease.

Later, the ancient Romans did much to develop systems for supplying clean water. Unfortunately, they also used lead piping, which probably inadvertently poisoned the populace and may have been one of the factors leading to the downfall of the Roman Empire. Even though such sanitary practices broke down in Europe during medieval and renaissance times, the need for these basic concepts is unchanged today.

During the 17th century, many cities began to construct water and sewer systems, and in 1793 a cholera epidemic struck Philadelphia, Pennsylvania, stimulating the construction of the first major water system in the United States.

Plumbing Codes

So, no matter how much shelter and comfort that beautifully constructed home of yours provides, you and your family are still highly dependent on the plumbing system. This is where a modern family of building regulations, such as the set that includes the *International Plumbing Code*® (IPC) and the *International Private Sewage Disposal Code*® (IPSDC), are essential in the design of acceptable sanitary systems. If the plumbing in a building isn't working correctly, it is unlikely that you're going to want to live there. The minimum provisions for plumbing systems result in an end product that will bring years of enjoyable and comfortable living.

Many of us have no intention of doing our own plumbing. We would rather enlist the services of a professional. Nonetheless, having a better understanding of the basic concepts of plumbing as outlined in the IPC will no doubt assist us in our understanding of what is being done and help us better communicate our needs and concerns to the plumber.

Where Did Modern Plumbing Codes and Their Requirements Come from Anyway?

The International Plumbing Code: A Guide for Use and Adoption (ICC, 1998) provides a very helpful overview of the technical basis of the IPC. Here are some highlights from its first chapter:

The IPC was the first code developed with the full cooperation of the three model code groups: Building Officials and Code Administrators International, Inc. (BOCA), International Conference of Building Officials (ICBO), and Southern Building Code Congress International, Inc. (SBCCI). The intent was to regulate plumbing with the most technically accurate code. The original intent of the IPC was to recognize all acceptable methods for the various plumbing systems. The code did not attempt to arbitrarily restrict any method, material, concept or system. Since its initial development, the IPC has been updated through an annual code change process with participation from nationally recognized industry experts.

Development of the Code

The IPC was developed through a cooperative effort of BOCA, ICBO and SBCCI (see Figure 1-2). The first draft of the IPC was prepared by a select group of plumbing officials working with the staff of the three organizations. This committee reviewed the contents of the BOCA *National Plumbing Code*, the ICBO Plumbing Code and the SBCCI *Standard Plumbing Code*.

A draft of the code was prepared for review by the industry. The first draft contained only excerpts from the three plumbing codes, with no new concepts or ideas added. All of the allowable practices were already permitted and used by one of the model plumbing codes. The premise was that the code should have its origins based on the content of the existing plumbing codes.

It was recognized that no one part of the United States had used all of the plumbing practices that would be permitted in the new plumbing code. Terminology would also be an initial difficulty since different parts of the country use the same name to describe different plumbing systems.

When the draft was issued, it was subject to a review process through a series of public hearings. A committee of plumbing officials was appointed to consider all of the testimony on the first draft. The hearings were well attended by plumbing experts from all areas of the industry. New provisions were proposed for inclusion at the hearing, including a complete rewrite of the backflow section. The new requirements received overwhelming support at the public hearings.

After modifying the draft to include the acceptable changes based on the testimony at the public hearings, the document was forwarded to the membership of BOCA, ICBO and SBCCI. The three organizations voted unanimously to accept the new code as a replacement document for their organizations' plumbing codes. In 1995, the International Code Council, Inc., published the first edition of the IPC.

Recognition of New Technology

The original committee that drafted the IPC developed a philosophy based on acceptance of new technology, including new materials and products, as well as new methods of installation.

While the acceptance of new technology was paramount to the IPC, any new idea, concept or material must be substantiated with technical documentation and reviewed through the open code change process.

Code Change Process

The IPC has an annual code change cycle for the review of all new proposals that is open and available to everyone. The ICC accepts code change proposals submitted before the deadline date, with no limitations placed on the submittal of code changes. Every code change is reviewed by the staffs of the three model code groups to address any administrative concerns. This provides every proponent of a code change with the best opportunity of being considered favorably.

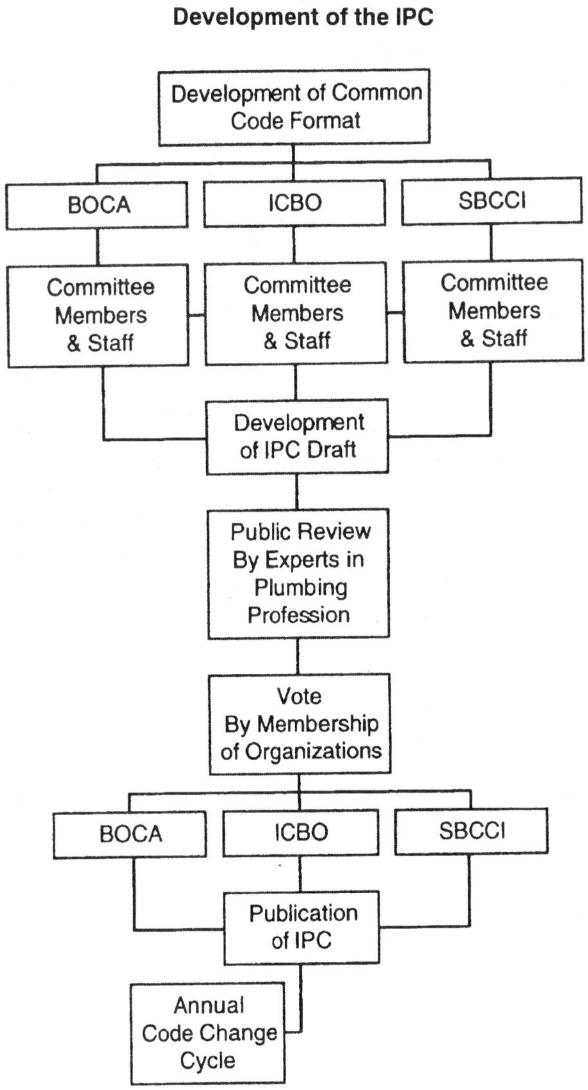

Development of the IPC

Figure 1-2 *Development of the IPC*

The code changes are published in a document for distribution to any interested party. A public hearing is scheduled for discussion by the proponents and opponents of each code change. The hearings are conducted before the ICC Plumbing Code Change Committee, a select group of plumbing professionals appointed by the sponsoring organizations: BOCA, ICBO and SBCCI. The committee is made up of plumbing inspectors, plumbing engineers, labor representatives and representatives of testing laboratories.

After hearing public testimony, the committee votes to recommend either approval, approval with modifications or disapproval of the code change. The results of the committee are published with reasons for every action taken.

The recommendations of the committee can be challenged during the second series of public hearings, where any challenged code change is open for discussion. The vote at the second hearing is by the voting membership of BOCA, ICBO and SBCCI. The membership can either agree or disagree with the committee's recommendation. A two-thirds majority of those voting is required to overturn the committee's recommendation and approve a code change.

The approved code changes are published in the supplement to, or the new edition of, the IPC.

Administration and Enforcement

Chapter 1 of the IPC follows the guidelines established by the legal community for the regulation of construction codes. The IPC is consistent with the recommendations of *Legal Aspects of Code Administration*. The administration requirements in the IPC recognize that once adopted by a jurisdiction, the code becomes a legal document. The administration and enforcement become the responsibility of the local jurisdiction. This is the philosophy regarding the adoption of any model code.

Alternative Approval

Section 105 of the IPC includes the requirements for alternative approval. This section, often considered the most powerful section of the code, follows the guidelines of the Federal Trade Commission for permitting the acceptance of new technology. It permits the code official to accept any alternative material, method or equipment that may not be recognized directly in the code.

The IPC is unique in specifying the requirements for alternative engineered design in the approval section. These provisions are consistent with the various state engineering and architectural registration acts. A registered design professional is permitted to design any plumbing system, provided that he or she has adequate technical documentation and testing to justify the alternative design. The alternative engineered design section was originally developed by Bernie McCarty, P.E., past president of the American Society of Plumbing Engineers (ASPE). McCarty submitted the code text on behalf of ASPE, in support to the Society's position regarding engineering design.

Consistent with Federal Guidelines

The IPC was developed consistent with federal guidelines regarding seismic protection and floodproofing. The seismic requirements in the IPC are consistent with the recommendations of the National Earthquake Hazards Reduction Program (NEHRP). The IPC references the building code for specific regulations relating to the location of the building.

The floodproofing requirements in the IPC were developed through a contract with the Federal Emergency Management Agency (FEMA). FEMA requested that references be located throughout the code to address the necessary floodproofing requirements.

Referenced Standards

The IPC relies on references to nationally developed consensus standards. To assist the code user, the IPC directly references the appropriate standard throughout the body of the code. The complete list of the referenced standards appears in Chapter 13, listed in order of the promulgating organization.

The ICC developed a criterion for the acceptance of referenced standards. To ensure fairness, the standards are required to be developed by the consensus process.

The standards are also required to be written in mandatory language without permissive or subjective text, allowing the standard to be a legally enforceable document. If there is any permissive text in a standard, it raises the issue of enforceability and who will make the decision regarding the permissive requirement.

All standards are reviewed for adherence to the ICC policy. What follows is a summary of where the technical provisions of the IPC came from. It is also excerpted from reference (ICBO, 1998).

Fixture Requirements

Minimum Number of Fixtures Required—The IPC specifies the minimum number of plumbing fixtures required for every building occupancy, based on both the number of building occupants and the occupancy classification. To avoid confusion, the table has been converted to values that are consistent with the building code occupant load tables.

Potty Parity—Around 1980, it was recognized that plumbing codes were providing an injustice to the female population by requiring an inordinate amount of plumbing fixtures for the male population. The inequity resulted from plumbing codes specifying a minimum number of water closets, as well as a minimum number of urinals, for the male population. It was also recognized that the waiting period for the female population far exceeded any waiting time incurred by the male population. Studies were performed by Dr. Sandra Rawls at the University of Virginia, Stevens Institute of Technology, the National Restaurant Association and the ASPE Research Foundations. The results of these studies are reflected in the IPC.

In developing code requirements for assembly buildings, the studies used the waiting time required. Hence, the urination process was analyzed for determining the minimum number of fixtures. If the female population requires twice as long to complete the urination process, they would need twice as many fixtures to have the same waiting period as the male population. Thus, the code reflects requirements for twice as many water closets in the ladies' room when compared to the men's room.

Studies by the National Restaurant Association indicated that "potty parity" was not required for restaurants or nightclubs. As a result, the average time factor for fixture use (rather than waiting time) can be applied, resulting in an equal distribution of the number of plumbing fixtures.

To prevent the inequality of fixtures from occurring in the future, the requirement for a urinal to be a mandatory fixture was removed from the code. This was also necessary for smaller toilet rooms that were designed for one fixture. The urinal is permitted to be substituted for a maximum of 50 percent of the required number of water closets. (This was changed in the 2000 IPC to 67 percent.)

Accessible Plumbing Fixtures—The IPC directly references the *International Building Code®*, which regulates accessible plumbing fixture requirements and identifies the requirements for accessibility for various buildings in Chapter 11.

Type B Dwelling Unit Accessibility—The IPC is the only model plumbing code to include specific requirements for Type B dwelling units. Type B dwelling units are designed to be readily adaptable to handicap access. The requirements have been coordinated with the Fair Housing Act.

Installation of Fixtures—The IPC regulates the spacing of plumbing fixtures to provide both comfort and social privacy. The spacing requirements are based on the results of a study conducted at Cornell University, published in a book entitled *The Bathroom*. Alexander Kira headed a study that completely analyzed the use of the various plumbing fixtures. The study concluded that adequate spacing was required between fixtures in public toilet rooms to help facilitate their use and to avoid direct body contact between users of fixtures.

The spacing for urinals reflects dimensions that are typically not followed in other plumbing codes; i.e., a 30-inch spacing. If a 24-inch spacing is provided, the result will be direct contact with the user of the adjacent fixture.

The study at Cornell was also used to determine the minimum size of a shower to accommodate movement of the individual to allow for the cleansing of the lower extremities.

Individual Fixture Requirements: An example would be the backflow requirements for fixtures and appliances. Section 406.2 requires an automatic clothes washer to have an integral air gap built into the machine, or the water supply must be protected against backflow in accordance with the requirements of Section 608. While residential and coin-operated automatic clothes washers have integral air gaps, this is not true for many large commercial machines. For these larger machines, backflow protection is provided on the water supply to the appliance. The backflow requirements are also similar for dishwashers. Many plumbing codes have a tendency to specify requirements for residential appliances and fixtures while leaving out the requirements for commercial equipment. The IPC is complete in specifying regulations for both residential and commercial appliances and fixtures.

Water Piping Systems Piping Material: The acceptable piping materials for water service and water distribution systems are listed in Tables 605.4 and 605.5. The IPC accepts all the common water piping materials, such as copper tubing, CPVC, galvanized steel, cross-linked polyethylene and PEX-AL-PEX. There are no arbitrary restrictions or prohibitions placed on the installation of any water piping materials. The tables identify the acceptable materials by reference to the ASTM or CSA standards. The IPC relies on these organizations for the development of acceptable material standards. Each standard is reviewed for completeness and compliance with the ICC standards policy.

Design Criteria for Sizing—Table 604.3 of the IPC specifies the minimum criteria for the design of a water distribution system, but is often misunderstood to be the minimum flow rates required for the specified plumbing fixtures. However, the criteria are used only for purposes of sizing a water distribution system. The values for flow rates and minimum pressures are used independent of one another. The flow rate values are used in determining the peak demand of the system, and the pressure values are the minimum requirements for the most demanding fixture operating under a peak demand condition. The flow rates are similar to the values published in the ASPE Data Book and are consistent with the fixture requirements specified in ANSI/ASME A112.18.1 and ANSI/ASME A112.19.6.

Maximum Flow Rates—The federal government imposed mandatory requirements for plumbing fixtures as part of the legislation for water and energy conservation. The IPC maintains consistency with the federal legislation by specifying the maximum flow rates (at specified pressures) that are the same as the federal legislation.

Minimum Pipe Size—Table 604.5 specifies the minimum water-pipe size required on the supply to each fixture. The minimum pipe sizes specified are consistent with those specified in the ASPE Data Book.

System Sizing Requirements—The IPC requires the water distribution system to be designed in accordance with accepted engineering practices. This provides the system designer with the flexibility to use any approved sizing method. The IPC permits the plumbing engineer to evaluate each water distribution system for peak demand and size the system accordingly. Plumbing engineers and system designers have been employing computer programs to size water distribution systems more accurately.

Many of the water-pipe sizing procedures use the concept of supply fixture units to determine the minimum pipe size. This method was originally developed by Dr. Roy B. Hunter, who wrote in BMS 799, "The design layout and the selection of material and pipe layout and the selection of material and pipe [should be] delegated to an engineer experienced in this field." The ASPE Research Foundation called for the abandonment of supply fixture-unit sizing methods in a report to the plumbing engineering community. The fixture-unit sizing method is considered out of date for properly sizing water-distribution systems.

While the IPC does not mandate a method that must be followed for sizing a water-distribution system, one method of sizing is provided in Appendix E. This method follows the precepts of Hunter's original water-pipe sizing procedure. Appendix E is provided for assistance to the code user, especially for smaller buildings that are designed by plumbing contractors.

Water Hammer—The plumbing community has recognized that water hammer in a piping system can only be controlled by preventing the occurrence, or with the installation of water-hammer arresters. Julius Ballanco, P.E., reported that controlling velocity to prevent water hammer was dependent on the type of water piping material installed. Section 604.9 of the IPC follows the engineering guidelines for water-hammer control. Both Ballanco and Steele reported that air chambers are ineffective in controlling water hammer. Both individuals reported that the only viable method for preventing the occurrence of water hammer was the installation of water-hammer arresters, or by controlling the velocity of flow in the piping.

Backflow Protection—Backflow protection is considered the most important aspect of a plumbing code. The backflow protection requirements in the IPC have been developed with the input of the leading backflow protection experts in the country using the latest information and references to national consensus standards.

Air Gap—The most common method of protecting the potable water supply is with an air gap. The minimum air-gap requirements are consistent with the air-gap requirements specified in ANSI/ASME A112.1.2 that were developed in 1942. They have been the mainstay of the plumbing industry for air-gap requirements.

Maintenance of Hot Water—In large buildings, the design professional will often design a water distribution system with long runs of piping for the hot water. The IPC requires the hot water temperature to be maintained in the piping to conserve both water and energy. The recognized methods of maintaining the water temperature are recirculation systems or temperature maintenance systems such as heat tapes. The maintenance of the temperature in hot water piping systems is consistent with the engineering practices of both ASPE and ASHRAE.

The IPC requires all water heaters to have pressure and temperature relief protection. The level of protection is consistent with the requirements of ASPE and ASHRAE.

Sanitary Drainage Systems

Piping Material—The acceptable piping materials for sanitary drainage systems are clearly listed in four tables: Tables 702.1, 702.2, 702.3 and 702.4. The IPC accepts all of the common drainage piping materials, such as ABS plastic pipe, cast-iron soil pipe, copper tubing, galvanized steel pipe, polyolefin pipe and PVC plastic pipe. There are no arbitrary restrictions placed on the installation of any of the piping materials.

One of the concerns with the installation of a sanitary drainage system is the impact the piping material has on the fire-protection aspects of a building. The building code requirements for pipe penetrations of required fire-resistive assemblies distinguishes between combustible piping materials, such as ABS and PVC, and noncombustible piping materials, such as cast-iron soil pipe and copper tubing. (All types of piping have listed assemblies to resist fire spread.)

Back-to-Back Water Closets—The flushing dynamics for a 1.6-gallon-per-flush water closet has had an impact on the use of double sanitary tee fittings. To avoid an interruption of flow, the IPC prohibits the installation of double sanitary tee fillings for back-to-back water closets. When water closets are installed back to back, either a double tee-wye or a double wye fitting must be used. (See Figure 4-3 in Chapter 4.)

Fixture Units—The fixture-unit sizing method in the IPC was developed by Dr. Hunter at the National Bureau of Standards. The concept was reported in BMS 65.19. In a companion report, SMS 66.20, the fixture-unit sizing method was codified. This report has served as the basis for the fixture-unit sizing method in all modern-day plumbing codes. Hunter's original concept was to assign every fixture a probability value for purposes of sizing a system to prevent uninterrupted flow.

Various plumbing codes have modified the table by adding extensive listings of numerous fixtures. The IPC maintains the original philosophy by listing the major category of fixtures. For example, a commercial kitchen sink would have a fixture-unit value of 2 drainage-fixture units (dfu) based on the classification of sink. Similarly, a mop sink in a janitor's closet would also have a fixture-unit value of 2 dfus under the same classification.

Hunter only distinguished between public and private fixtures in the listing of water closets and bathroom groups, and the IPC maintains that notion. Since the publication of BMS 66 in 1940, there have been changes to the majority of fixture-unit values. These changes resulted from additional work using Hunter's methods.

With the introduction of 1.6-gallon-per-flush (gpf) water closets, there have been various research evaluations to determine if the fixture unit table should be adjusted for low-consumption water closets. In a paper delivered by Dr. Larry Galowin and Professor John Swaffield it was reported that the dfu method for low-consumption water closets may cause a problem because the fixtures have a higher peak flow rate (although for a shorter period of time). Galowin and Swaffield reported the need for having smaller pipe sizes for the discharge of 1.6 gpf water closets.

Rather than adjusting the fixture-unit value, the IPC permits manufacturers to have their products individually tested for establishing a low fixture unit value. This is necessary with the different flushing performance of the various manufacturers' water closets.

Three-inch Limitation for Water Closets—Although many other plumbing codes have a limitation, the IPC has not limited the number of water closets permitted on a 3-inch drain. Dr. Hunter, in BMS 5 and BMS 66, never placed any limitation on the number of water closets on a 3-inch drain. His sizing method was based solely on the fixture-unit values. The limitations on the 3-inch drains were added to plumbing codes long after Hunter's death, in what has been identified as a misinterpretation of his original research.

In a report by Galowin and Swaffield, and later by Galowin, Campbell and Swaffield, the arbitrary limitation on the number of water closets permitted to connect to a 3-inch drain was denounced. The IPC has chosen to base the code requirements on the latest technical information; hence, there is no limitation. The sizing is based on fixture-unit value only.

Sizing of Drainage System—The drainage pipe sizing method in the IPC also follows Hunter's methods. Hunter developed drainage pipe sizing tables based on fixture-unit values in his research. The tables were modified based on follow-up research published in NBS Monograph 31.24.

Sizing of Offsets in Stacks—An offset in a drainage stack has special sizing requirements in the IPC. The sizing method is based on the procedures originally specified in BMS 66. The offset is permitted to have a greater amount of flow than a connecting horizontal branch because it is considered a part of the stack. The sizing is the same as the sizing for a building drain.

Pitch of Drain Pipe—The minimum pitch required for a drainage pipe is specified in Table 704.1. The IPC permits a 3-inch drain to be pitched $1/8$ inch per foot. Other plumbing codes require 3-inch drains to be pitched a minimum of $1/4$ inch per foot. The discrepancy arises from the calculated velocity of flow in a drain using the Manning Expression. When using the proper roughness value, 3-inch pipe can be pitched at $1/8$ inch per foot and meet the minimum velocity requirements.

It should be noted that Hunter specified a minimum pitch of $1/8$ inch per foot in BMS 66 and the ASA A40.8 *(National Plumbing Code, 1955)* also permitted a pitch of $1/8$ inch per foot for a 3-inch drain.

Computerized Sizing Methods—The IPC allows the drainage system to be designed by computerized method. This method of sizing is more accurate than the method originally developed by Hunter.

There are two computer programs recognized for sizing drainage systems. One program was developed at the National Bureau of Standards. The other program is a complete system modeling program developed at Heriot-Watt University in Edinburgh, Scotland, and introduced to the plumbing engineering community at the 1996 ASPE Convention.

Storm drainage requirements are consistent with the engineering guidelines of the American Society of Plumbing Engineers and the American Society of Civil Engineers.

Cleanouts—The requirements for cleanouts have changed over the years with the improvement of drain-cleaning equipment. The IPC bases the cleanout requirements on using modern-day drain-cleaning equipment (see Figure 1-3).

Sewage Pumps and Ejectors—The IPC follows the recommendations of the Sewage and Sump Pump Manufacturers Association for the design and installation of sewage pumps and ejectors. The sizing of the drain pipe from a sewage pump is based on full flow with a minimum velocity of 2 feet per second.

The IPC does not arbitrarily require the installation of dual pumping equipment for sewage sumps. The use of this equipment is a decision for the building owner or designer to make. The added expense of dual pumping equipment (more than double the price because of the required controls) does not provide any additional protection of public health or safety.

Grease Interceptors—Grease interceptors are required for all restaurants, commercial kitchens and

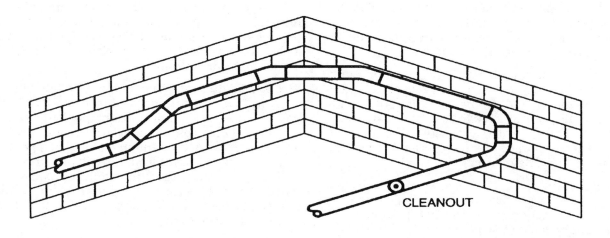

Figure 1-3 *Horizontal piping arrangements*

similar food-handling establishments. The grease interceptor separates the grease before discharging to the public sewer, thus protecting the sewage treatment system.

The IPC is one of the few plumbing codes to permit food-waste grinders, which are the largest sources of grease, to discharge to a grease interceptor. A study in Wisconsin discovered that large volumes of grease were not being intercepted in commercial kitchens because the grease was flushed down the food-waste grinder during the initial washing of the dishes.

Venting Systems

Acceptable Venting Methods—The most common form of venting is that which serves only one fixture trap. The vent pipe allows air to enter as well as relieve any pressure that might be experienced in the drainage system. This method of venting has been included in every plumbing code since the advent of modern indoor plumbing. While an individual vent is considered the easiest method of protecting the trap, it also results in the most extensive amount of piping.

Common Venting—Common venting is considered to be a form of individual venting. The common vent serves two fixtures that are located in the same general area. Common venting was recognized by Hunter in SMS 66.26. Offset common venting, including the piping arrangement and sizing, was reported in BMS 119.

Wet Venting—Wet venting is a system that combines the venting of fixtures in a bathroom within a dwelling unit. The system can be extended to include all the fixtures located in two adjacent bathrooms. Hunter first reported on wet venting, including the concept, in SMS 66. It was further investigated by French, Eaton and Wiley.

The wet-venting systems permitted by the IPC are also recognized in the *ASPE Data Book* as an acceptable design. Wet venting is also addressed in numerous other widely recognized plumbing design manuals, including *Engineered Plumbing Design* and *Practical Plumbing Engineering*.

The sizing of the wet vent is consistent with the sizing determined through research at the National Bureau of Standards. Additional testing at Stevens Institute of Technology verified the sizing of the wet vent piping.

Circuit Venting—Circuit venting is an arrangement that permits up to eight fixtures to be protected with a single vent pipe. The requirements are very specific to ensure the performance of the system.

The system performs well because the drainage flow is on a horizontal plan and siphon action from the horizontal flow is minimal. Extensive research into the performance of circuit vented systems was conducted at the State University of Iowa. The research concluded that the single vent for the eight fixtures provided the necessary protection of the trap seal.

Circuit venting was included in Hunter's research and reported in BMS 66. This venting method has long been recognized by the plumbing community and is included in the *ASPE Data Book*.

The combination drain and vent system is based on the same premise as the circuit-vented system. Most plumbing codes place arbitrary restrictions on combination drain (waste) and vent systems because the systems appear too good to be true.

The combination drain and vent is a method of venting sinks, lavatories, floor drains and standpipes located on the same floor. The main drain pipe must be oversized of the intended discharge load. This reduces the siphonic action and prevents the development of any pressure condition in the piping. A single vent must connect to the main drain.

The performance of the combination drain and vent system was verified in tests conducted at Stevens Institute of Technology. If sized according to the table in the IPC, the study concluded that the distance from a trap to a vent does not have to be limited in length.

Waste Stack Venting—There have been many types of single-stack plumbing systems used throughout the world. One of the most common systems used in the United States is the waste stack vent. This venting method allows fixtures other than water closets and urinals to connect directly to a stack without any additional venting. The performance of the system is based on the oversizing (or underloading) of the drainage stack. Testing of stack venting arrangements was performed by French and the National Bureau of Standards.

A waste stack vent system also requires every fixture to connect independently to the stack. This prevents interference of flow from one fixture to another. The system is included in the *ASPE Data Book* as an acceptable venting system.

Vent Pipe Sizing—The vent pipe sizing requirements in the IPC are based on the latest technical information in studies conducted at the National Bureau of Standards. Table 916.1, the main venting table, is derived from Wyly and Eaton's research, published in NBS Monograph 31. This was one of the most conclusive studies on the air-movement requirements in a plumbing drainage and vent system. Most modern plumbing codes base their drainage stack sizing and vent stack sizing on the results published in this report.

Air-admittance Valves—The IPC accepts the use of air-admittance valves for vent terminals of individual and branch vents. These devices are regulated by a nationally recognized consensus standard. There has been extensive research into the performance of air-admittance valves. Swaffield and Galowin, whose studies validated the performance of air-admittance valves, reported on this in their engineering design book (see Chapter 2 for illustration).

Air-admittance valves have been the subject of criticism from the time they were introduced to the plumbing industry in 1988. The concerns range from placing faith in a mechanical device to the amount of material and labor saved through their use. However, Swaffield and Campbell reported that the use of air-admittance valves improves the performance of a drainage system by controlling the amount of air introduced into the system.

Administrative Details

Purpose of the International Plumbing Code

The plumber, the general contractor for whom he or she works, and the jurisdictional inspector are responsible for making sure that your plumbing system is installed correctly and is adequate for its purpose. Does this mean that expensive and more than adequate standards are required? The answer is no.

Notice that IPC Section 101.3, Intent, states:

"The purpose of this code is to provide minimum standards to safeguard life or limb, health, property and public welfare by regulating and controlling the design, construction, installation, quality of materials, location, operation, and maintenance or use of plumbing equipment and systems."

By "minimum standards" the code means minimum acceptable levels to safeguard health and safety. Of course, "minimum acceptable level" is subject to opinion, and the code defines nationally recognized standards to which the product or design must adhere. Without a viable code development consensus process, such as described previously, and prescribed scientific test and engineering criteria, regulations could be affected by vested interests.

Speaking of minimums, of course, carrying a shovel and going out to the back 40 to "take care of business" could be construed as meeting this intent (and may even result in some level of localized disease control). But, nowadays, it has been proven a more efficient and certainly a more healthy practice to connect local plumbing to a community-wide sanitary system.

This is not to say that more rural or less developed areas can't use private sewage systems. That is the reason for the *International Private Sewage Disposal Code*. We will discuss this code in Chapter 8.

Applicability

In the construction of a new building, the IPC will assist in determining the types of materials to be used, the acceptable methods of construction and other requirements. The IPC applies not only to the construction of new buildings, but also to the maintenance, repair and remodeling of existing buildings.

In maintaining an existing system, you need to be sure that new installations will continue the existing system's (presumably) smooth operation. If they do not, even if the system does not completely fail, you will have a problem. It may result in an unsafe condition of one's property or health.

Care must also be exercised when the building's use or occupancy has changed. When a building or portion of a building was previously used for a specific purpose and will now be put to a different use, the existing design may be inadequate. For instance, the sewage system could be overwhelmed by increased sewage or water demands. Here, again, the IPC will assist you in determining what the proper type of installation for the new use would be.

Another instance of concern occurs when buildings are moved from one location to another. They may be moved within the same community or moved to a completely different jurisdiction. The new plumbing system's site-specific parameters must be examined to ensure that they meet the intended purpose and will perform adequately.

Building or Plumbing Departments

Obviously, we have all determined that we don't want to carry a shovel around for the rest of our days. We want to have a working plumbing system that will make it comfortable for us to "take care of business" quietly and away from the public. But there is someone else who wishes to have you "take care of business" properly— and that is the code official. (By the way, when the code uses the term "code official" it refers to the designated manager of the department and all the employees under that individual's sphere of responsibility).

In the vast majority of jurisdictions, the code official is the person who manages the Building Department. In fact, this person may be the only employee in the department in smaller jurisdictions, or the manager of hundreds of people in larger ones. In some cases, this official may be the Director of Community Development or the Director of the Public Works Department or even the Fire Marshal.

The plumbing code explains how this code official is there to assist you. Section 104 outlines the duties and powers of the code official:

"The code official shall enforce all of the provisions of this code and shall act on any question relative to the installation, alteration, repair, maintenance or operation of all plumbing systems."

So the code official and departmental employees have to be knowledgeable of plumbing installation and the plumbing code (and how the code interrelates with, and affects, other International Code regulations). These public employees will be an asset in helping you accomplish your goal of installing a safe and effective plumbing system. We'll discuss applications for permits, the permits themselves, plan review and inspections a little bit later. The code official and departmental personnel are responsible for all these activities. We cannot overemphasize that these departmental personnel are there to assist. It may seem that the process is overly bureaucratic and legalistic, but the administration of building regulations is a police power of the community and needs to be structured in that manner.

Permits

There is, however, one thing that will link you and the code official together for sure. Something in common that you'll both be using and looking for —the ever-popular plumbing permit. Why do I need a permit? you may ask. Well, one reason is that the code official would like to eat. The code official expects to be paid for work that he or she is doing. But, obviously, that's not the real reason for the permit. The primary reason for the plumbing permit is to ensure that you and the code official are working together to install your plumbing system properly within sanitary guidelines.

You must first find out from your code official, and from the IPC, what type of work is going to require a permit. Section 106.2 tells us what work is exempt from a permit. And, of course, you'll want to talk with the code official in your jurisdiction to find out if there are other matters regarding permits and permit exemptions.

There is one portion of Section 106.2 that you need to keep in mind. That is:

"Exemption from the permit requirements of this code shall not be deemed to grant authorization for any work to be done in violation of the provisions of this code or any other laws or ordinances of this jurisdiction."

Remember, no matter how small the installation or repair, if it's not done to code, it is not only nonconforming, but will likely fail to work adequately and safely.

You should expect to pay a fee with the application for a plumbing permit. Here, again, you will need to find out from the code official of the jurisdiction what the fee will be.

In Chapter 3 we'll discuss what it takes to get started with your plumbing installation. There we will highlight more about what items are needed in order to purchase the plumbing permit and how you and the code official will work together in accomplishing the goal of a safe, properly installed plumbing system.

To get started, you need to know the language. Plumbing has a unique language, so understanding the meaning of certain terms will help you to accomplish your goal.

So the next time someone says to you, "Plumbing codes, who needs them?" we hope you will say, "I do!"

References:

1. O'Bannon, Robert E., *Building Department Administration.* Authorized Second Edition, ICBO, 1989.
2. ICC, *International Plumbing Code: A Guide for Use and Adoption*, copyright 1998, ICC.

2

Definitions

Overview

The arcane language of plumbing is no different than the jargon used in any other trade or profession. Unless a person is trained or engrossed in the field, it will be unfamiliar. It takes time and usage to perfect any language, but when it is separated into "bite-size" pieces, eventually it will be understood and can be communicated to others. Even those in the trade don't completely understand *all* the terms.

Mastery of any profession includes having a fundamental understanding of the trade's terms. From there, skill accumulates and understanding deepens. The definitions listed below help in comprehending the system requirements as well as the legally specific code requirements.

This chapter contains the complete list of 2000 IPC definitions and some illustrations to augment them. It is hoped that there is enough information to aid in the design and application of plumbing systems. Also included are other definitions that will assist in understanding some concepts.

Not all terms used in the plumbing industry are included. Terms left out include definitions with reference to tools, specific fittings, chemicals used in the industry, auxiliary terms not directly associated with the trade and engineering terms. These terms can be found easily by contacting toolmakers or supply houses for tools, manufacturers of chemical products, trade reference books, the dictionary for general terms, and other various texts or resources on any given subject area.

ABS. See *Acrylonitrile-Butadiene-Styrene.*

Absorption Trench. A trench bedded with clean, coarse aggregate around a distribution pipe that is covered with a permeable fabric under a 12-inch minimum earth cover; e.g., septic fields.

Accepted Engineering Practice. That which conforms to accepted principals, tests or standards of nationally recognized technical or scientific authorities.

Access Cover. A removable plate, usually secured by screws, to permit access to piping, controls or hidden equipment for the purpose of inspection, repair or cleaning.

Access Door. An operable, hinged door mounted in framework for access to piping, controls or equipment for the purpose of inspection, repair or cleaning. This type of access is more user-friendly.

Accessible. Having access to fixtures, equipment, appliances or required fittings by means of removing a panel, plate or operable door. See *Ready Access*.

Acid Waste Piping. System piping and fittings that are resistant to acid contents usually found in manufacturing processes.

Acrylonitrile-Butadiene-Styrene (ABS). A plastic generally used for drainage systems.

Adapter Fitting. An approved connecting device that suitably and properly joins or adjusts pipes and fittings that do not otherwise fit.

Administrative Authority. See *Code Official*.

Aerator. A device used to mix air and water with water distribution and water conservation in mind; e.g., faucets.

Air Break. A piping arrangement in which a drain from a fixture, appliance or equipment discharges indirectly into another fixture, receptacle or interceptor at a point below the flood level rim. (See Figure 2-2.)

Air Gap (Drainage System). The unobstructed vertical distance through the free atmosphere between the outlet of a waste pipe and the flood level rim of the receptacle into which the waste pipe is discharging. (See Figure 2-2.)

Air Gap (Water Distribution System). The unobstructed vertical distance through the free atmosphere between the lowest opening from any pipe or faucet supplying water to a tank, plumbing fixture or other device and the flood-level rim of the receptacle. (See Figure 2-3.)

Air Test. Test performed on the plumbing system to ensure its integrity. It is administered after the piping system(s) are complete.

Air-admittance Valve. One-way valve designed to allow air into a plumbing drainage system on demand from area fixtures. Gravity seals the valve after use and prevents sewer gases from escaping. (See Figure 2-1.)

Alternative Engineered Design. A plumbing system that performs in accordance with the intent of the IPC and provides for the protection of public health, safety and welfare.

Anchors. Devices that attach supports, equipment and fixtures in place.

Angle Stop. A 90-degree water supply valve at its termination point where it connects to the fixture or equipment.

Angle Valve. A 90-degree valve in a water distribution system that allows for that change in direction.

Anneal. The process of heating piping in order to soften it; e.g., soft copper.

Anode. A magnesium rod found in water heaters to prevent corrosion by chemical properties.

Antisiphon. A term applied to valves or mechanical devices that eliminate siphonage.

Approved. Approved by the code official or other authority having jurisdiction.

Area Drain. A receptacle designed to collect surface or rain water from an open area.

Asbestos. A silicate of calcium and magnesium, usually occurring in fibrous material. Pipe covering and cloth used to be made from this material because of its insulating properties. No longer used because of its effect on health.

Asphaltic Solution. A black/brown bituminous mixture used as a sealant or protectant. A by-product of tar-coal refining.

Aspirator. A fitting or device supplied with water or other fluid under positive pressure that passes through an integral orifice or constriction, causing a vacuum. Aspirators are also referred to as suction apparatus, and are similar in operation to an ejector.

Backfill. Clean fill material used to replace excavated dirt when covering up piping in the underground stages of construction.

Backflow

Backpressure. Pressure created by any means in the water distribution system, which, by being in excess of the pressure in the water supply mains, causes a potential backflow condition.

Backsiphonage. The backflow of potentially contaminated water into the potable water system as a result of negative pressure in the water system.

Backwater Valve. A device installed in the building drain or sewer where the sewer is subject to backflow that prevents drainage or waste from backing up into a low level or fixtures and causing a flooding condition.

Definitions

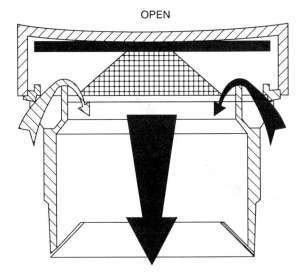
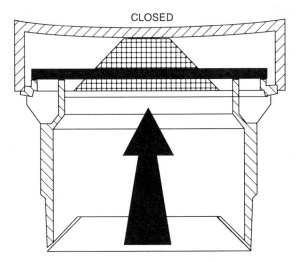

Figure 2-1 *Air-admittance valve*

Baffle. A separation wall used to restrict flows of air, gas, liquids or flue gases.

Ball Cock. A valve opened and closed by the fall or rise of a ball floating on the surface of water, whose elevation is controlled wholly or in part by the valve.

Basin. A receptacle used to hold water or other liquids for washing or other treatment.

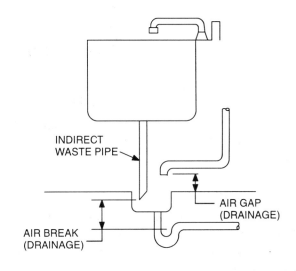

Figure 2-2 *Air gap, Air break*

Drainage. A reversal of flow in the drainage system.

Water Supply System. The backflow of water or other liquids into the water distribution system, thereby contaminating the potable water supply.

Backflow Connection. Any arrangement whereby backflow is possible.

Backflow Preventer. A device or means to prevent backflow.

Bacteria (Aerobic). Bacteria commonly found in septic type tanks that breaks down matter containing oxygen.

Bathroom Group. A group of fixtures consisting of a water closet, lavatory, bathtub or shower, including or excluding a bidet, an emergency floor drain, or both. Such fixtures are located together on the same floor level.

Battery Fixtures. Any group of two or more similar fixtures discharging into a common horizontal drain.

Beam Clamp. A device used to attach supporting hardware to beams.

Bedpan Washer. A device supplied with hot and cold water, or steam at times, used to clean or disinfect bedpans and related items. The contents are then flushed into the sanitary system.

Bell and Spigot Joint. Cast-iron piping system comprising a bell that receives pipe inside and is sealed using lead poured over packed oakum or rubber seal gasket.

Bidet. A fixture similar in stature to a water closet using hot and cold water to wash or bathe the genitals or posterior parts of the body.

Black Iron Pipe. Steel pipe either uncoated or protected in its black color state typically used for fuel gas piping.

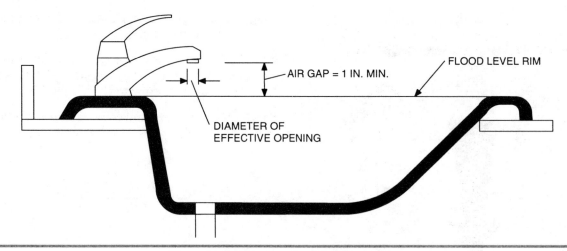

Figure 2-3 *Air gap for a lavatory faucet*

Boiler Drain. A valve located at the bottom of the boiler or water heater that is used to drain the equipment it serves.

Bonnet. Top nut that holds packing in place on a valve for sealant purposes.

Branch. Any part of the piping system except a riser, main or stack. (See Figure 2-9.)

Branch Interval. A distance, not less than 8 feet, along a soil or waste pipe corresponding in general to a story height, within which the horizontal branches from one story of a structure are connected to the stack. (See Figure 2-9.)

Branch Vent. A vent connecting to one or more individual vents with a vent stack or stack vent. (See Figure 2-9.)

Brazing. High temperature soldering with alloys that bond to metals in excess of 80°F.

Btu. British thermal unit.

Building. Any structure occupied or intended for supporting or sheltering any occupancy.

Building Drain. That part of the drainage system that extends from the end of the building drainage system, that receives the discharge from soil, waste, and other drainage pipes inside, and that extends 30 inches beyond the walls of the building and conveys the drainage to the sewer. (See Figure 2-9.)

Building Sewer. That part of the drainage system that extends from the end of the building drain and conveys the discharge to a public sewer, private sewer, individual sewage disposal system or other point of disposal.

Building Subdrain. That portion of a drainage system that does not drain by gravity into the building sewer. (See Figure 2-9.)

Building Trap. A device, fitting or assembly of fittings installed in the building drain to prevent circulation of air between the building drainage system and the building sewer.

Bushing. A short pipe fitting that is threaded on the inside and outside.

Bypass. A piping system that allows for a secondary supply for purposes of addition, repair or removal of valves, equipment or components of the primary supply piping.

Cap. A fitting that is screwed, glued or soldered to a termination point of any piping system.

Capillary Action. The process of drawing fluid into a space (e.g., heat drawing solder into a fitting).

Carrier. A mounting device that is attached to a structure that holds water closets, urinals, lavatories or sinks in place.

Cast Iron. Iron, carbon and silicon made by casting into a mold and producing pipe, fixtures and similar hardened products.

Catch Basin. A watertight sump used to collect sediment or oils from surface areas.

Cesspool. A pit, usually lined, that lets the liquid seep into the ground while holding the solids from the building drain (usually not permitted).

Check Valve. A one-direction valve that allows flow one way, the check operated by gravity or spring.

Circuit Vent. A vent connected to a horizontal drainage branch that vents two traps (to a maximum of eight traps) or trapped fixtures connected to a battery.

Cistern. A small covered tank for storing water. Generally, this tank stores rainwater to be used for purposes other than the potable water supply.

Clarifier. An interceptor or part of interceptor tank series.

Cleanout. An access opening in the drainage system used for the removal of obstructions. Types of cleanouts also include a removable plug or cap, or a removable fixture or fixture trap.

Closed System. A water piping system that is isolated by means of check valves, pressure regulating devices or backflow preventers that prevents the flow of water back into the source supply.

Closet Flange. Floor or wall flange that is screwed or bolted to structure with noncorrosive fasteners. The flange itself is part of the waste system.

Closet Gasket. A wax ring or sealant material used to form a watertight seal between the water closet bowl and the closet flange to which it is attached.

Code. Regulations, subsequent amendments thereto, or any emergency rule or regulation that the administrative authority having jurisdiction has lawfully adopted.

Code Official. The officer or other designated authority charged with the administration and enforcement of this code, or a duly authorized representative.

Combination Fixture. A fixture containing two or more compartments such as a laundry sink or three-compartment sink.

Combination Waste and Vent System. A specially designed system of waste piping embodying the horizontal wet venting of one or more sinks or floor drains by means of a common waste and vent pipe adequately sized to provide free movement of air above the flow line of the drain.

Combined. A building sewer that conveys both sewage and storm water.

Common Vent. A vent connecting at the junction of two fixture drains or to a fixture branch and serving as a vent for both fixtures.

Condensate Drain. A drain from a HVAC unit or condenser that is piped indirectly to the drainage system or other approved location.

Conductor. A pipe inside the building that conveys storm water from the roof to a storm or combined building drain. (See Figure 2-9.)

Confined Space. A room or space having a volume of less than 50 cubic feet per 1000 btu of the total combined ratings of all fuel-burning equipment in that space.

Contamination. An impairment of the quality of the potable water that creates an actual hazard to the public health through poisoning or through the spread of disease by sewage, industrial fluids or waste.

Continuous Waste. A series of connections to the drain outlets of a combination fixture that uses a common trap. (See Figure 2-4.)

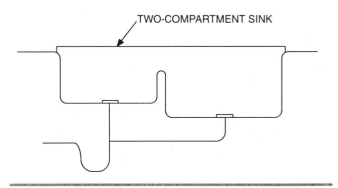

Figure 2-4 *Continuous waste*

Corporation Cock. A stopcock attached to a water main in the street that will be the first connection to the dedicated water meter.

Coupling. A female pipe-fitting used to join pipe in a straight line.

Critical Level (C-L). An elevation reference point that determines the minimum height at which a backflow preventer or vacuum breaker is installed above the flood-level rim of the fixture or receptor served by the device. The critical level is the elevation level below which there is a potential for backflow to occur. If the critical level marking is not indicated on the device, the bottom of that device shall be the critical level.

Cross Connection. Any physical connection or arrangement between two otherwise separate piping systems, one of which contains potable water and the other either water of unknown or questionable safety or steam, gas or chemical, whereby there exists the possibility for flow from one system to the other, with the direction of flow depending on the pressure differential between the two systems.

Crown Weir. The highest part of the trap before it flows horizontally to the drain. (See Figure 2-5.)

Dead End. A branch leading from a soil, waste or vent pipe; a building drain; or a building sewer terminating at a developed length of 2 feet or more by means of a plug, cap or other closed fitting.

Depth of Water Seal. The depth of water that would have to be removed from a full trap before air could pass through the trap.

Developed Length. The length of a pipeline measured along the center line of the pipe and fittings.

Dewpoint. The temperature of a liquid or gas at which condensation or evaporation occurs.

Dip of Trap. The top portion of the pipe at the bottom-most portion of the weir. (See Figure 2-5.)

Contractor's Guide to the Plumbing Code

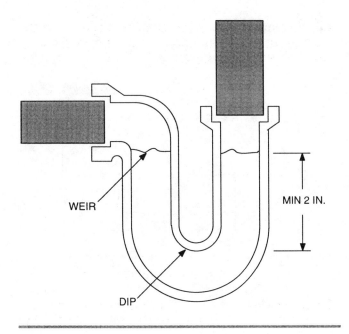

Figure 2-5 *Trap*

Discharge Pipe. A pipe that conveys the discharges from plumbing fixtures or appliances.

Distribution System. Part of the septic system, comprising a distribution box and leach lines, that disposes of effluent on the downstream side of the septic tank.

Downspout. An opening from a gutter system conveying rainwater to the building storm drain, combined building sewer or other disposal system away from the building.

Drain. Any pipe that carries waste water or waterborne wastes in the building drainage system. (See Figure 2-9.)

Drainage Fittings. A special type of fitting or fittings utilized in the drainage system. Drainage fittings are similar to cast-iron fittings, except that instead of having a bell and spigot, drainage fittings are recessed and tapped to eliminate ridges on the inside of the installed pipe.

Drainage System. All the piping within public or private premises that conveys sewage, rain water or other liquid wastes to a point of disposal. A drainage system does not include the mains of public sewer systems or a private or public sewage treatment or disposal plant.

    Building Gravity. A drainage system that drains by gravity into the building sewer.

    Sanitary. A drainage system that carries sewage and excludes storm, surface and ground water.

    Storm. A drainage system that carries rainwater, condensate, cooling water or similar liquid wastes.

Drainage-fixture Unit (dfu). A measure of the probable discharge into the drainage system by various types of plumbing fixtures. The drainage fixture value for a particular fixture depends on the volume rate of its drainage discharge, on the time duration of a single drainage operation and on the average time between successive operations.

Dry Well. A pit constructed to receive waste water and filled with gravel or coarse sand to aid in seeping into the soil.

Ductile Iron. Thick-walled iron piping and fittings usually cast in place—commonly used for water mains.

Effective Opening. The minimum cross-sectional area at the point of water-supply discharge, measured or expressed in terms of a circle or, if the opening is not circular, the diameter of a circle of equivalent cross-sectional area. For faucets and similar fittings, the effective opening shall be measured at the smallest orifice in the fitting body or in the supply piping to the fitting.

Effluent. Waste water, either industrial or sewage.

Ejector. Pump or device used to remove water or waste water from sumps or interceptors to a point where it flows freely away from the building drain. (See Figure 2-9.)

Electrolysis. Corrosion caused as a result of electric transfer of materials.

Emergency Floor Drain. A floor drain that does not receive the discharge of any drain or indirect waste pipe, and that protects against damage from accidental spills, fixture overflows and leakage.

Enamel. Smooth, protective covering over metals or pottery, usually heat applied.

Escutcheon. A wall or floor flange used on a pipe to seal openings and enhance the appearance of the valve or device served.

Essentially Nontoxic Transfer Fluids. Fluids having a Gosselin rating of 1, including propylene glycol, mineral oil, polydimethylsiloxane, cloroflourocarbon, hydrochloroflourocarbon, and hydroflourocarbon refrigerants; and FDA-approved boiler water additives for steam boilers.

Essentially Toxic Transfer Fluids. Soil, waste or gray water, and fluids having a Gosselin rating of 2 or more, including ethylene glycol, hydrocarbon oils, ammonia refrigerants and hydrazine.

Existing Installations. Any plumbing system regulated by this code that was legally installed prior to the effective date of this code, or for which a permit to install has been issued.

Expansion Joint. A fitting or fittings that allow for the thermal expansion of piping in a given area without compromising the integrity of that system.

Faucet. A valve end of a water pipe through which water is drawn from or held within the pipe.

Ferrule. A convex bushing used to compress over tubing to form a watertight seal. Also used with a tubing stiffener inserted in softer plastic tubing.

Filter System. Comprises filter media, fibrous material for separation of particulate and, at times, taste enhancers. Usually located near dispensing outlets.

Fitting. A part of a piping system that connects two or more pieces that serves a desired location or equipment. A fitting can change piping direction, reduce or increase size, and branch off.

Fixture. See *Plumbing Fixture.*

Fixture Branch. A drain serving two or more fixtures that discharges to another drain or to a stack. (See Figure 2-9.)

Fixture Drain. The drain from the trap of a fixture to a junction with any other drain pipe. Also known as a trap arm.

Fixture Fitting

   Supply fitting. A fitting that controls the volume and/or directional flow of water and is either attached to or accessible from a fixture, or is used with an open or atmospheric discharge.

   Waste fitting. A combination of components that conveys the sanitary waste from the outlet of a fixture to the connection to the sanitary drainage system.

Fixture Supply. The water supply pipe connecting a fixture to a branch water supply pipe or directly to a main water supply pipe.

Flaring System. A system whereby tubing is "flared" by means of a flaring tool that presses a cone-type press into tubing held in place by a split block assembly.

Flash Point. The temperature at which a liquid evaporates in sufficient quantity to ignite when a flame is applied.

Flashing. A waterproof assembly that is put around piping or venting and roofing material that seals either completely around or covers the top-most section.

Flood Zones

   Flood-hazard Zone (A Zone). Areas that have been determined to be prone to flooding but not subject to high-velocity waters or wave action.

   High-hazard Zone (V Zone). Areas of tidal influence that have been determined to be subject to wave heights in excess of 3 feet or subject to high-velocity wave run-up or wave-induced erosion.

Flood-level Rim. The edge of the receptacle from which water overflows.

Floor Sink. An indirect waste receptor, usually square, with grating, and made in many depths and widths.

Flow Pressure. The pressure in the water-supply pipe near the faucet or water outlet while the faucet or water outlet is wide open and flowing.

Flue. The ducting material that connects to appliances or equipment in which hot combustion flue gases flow to open atmosphere.

Flush Tank. A tank designed with a ball cock and flush valve to flush the contents of the bowl or usable portion of the fixture.

Flushometer Tank. A device integrated within an air-accumulator vessel that is designed to discharge a predetermined amount of water to fixtures for flushing purposes.

Flushometer Valve. A valve attached to a pressurized water-supply pipe and so designed that when activated it opens the line for direct flow into the fixture at a rate and quantity to operate the fixture properly, and then gradually closes to reseal fixture traps and avoid water hammer.

Foot Valve. A foot-operated valve usually found in clinical settings.

Free Circulation. The amount of air within the drainage system that ensures a safe margin for fluids to flow without back siphonage or broken trap seals taking place.

French Drain. A ditch filled with crushed rock or stones and covered with earth for yard drainage.

Funnel Fitting. A funnel-shaped fitting installed on the flat floor drain strainer, designed to receive intermittent or small quantities of indirect waste.

Galvanized iron. Zinc-coated iron that is used for rust inhibition.

Gate Valve. A fitting that controls the system flow by lowering or raising the "gate" part of the through valve, thereby opening or sealing the piping line.

Globe Valve. A stem and seat style valve that channels the flow in an "s" pattern as part of its opening and closing mechanism.

Grade. The fall or slope of drainage piping as compared to level (e.g., $1/4$ inch per foot would mean that for every foot measured, a rise or fall of $1/4$ inch should occur).

Grease Interceptor. A passive interceptor having a rated flow exceeding 50 gpm (189 L/m) and that is located outside the building.

Contractor's Guide to the Plumbing Code

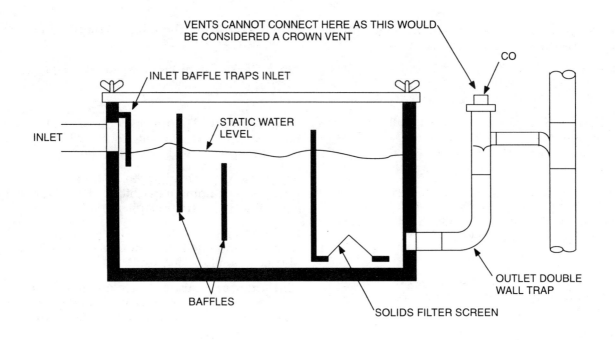

Figure 2-6 *Typical grease trap design*

Grease Trap. A passive interceptor having a rated flow of 50 gpm (189 L/m) or less and that is located inside the building (see Figure 2-6).

Ground Joint Union. A three-piece fitting that connects together by means of a machined face and counterface, to form a seal when the outer nut is tightened.

Groundwork. The plumbing that lies beneath the concrete slab or within the ground and is tested prior to structure covering.

Gutter. A roof channel system to receive rain water and direct it to leaders or downspouts.

Handhole Plug. A plug that serves fixtures or equipment that, when removed, acts to service fixtures or equipment by removing debris and cleaning that area.

Hangers. See *Supports*.

Head Pressure. Liquid or other pressure exerted downward as a result of vertical elevation.

Horizontal Branch Drain. A drainage branch pipe extending laterally from a soil or waste stack or building drain, with or without vertical sections or branches, that receives the discharge from two or more fixture drains or branches and conducts the discharge to the soil or waste stack or to the building drain. (See Figure 2-9.)

Horizontal Pipe. Any pipe or fitting that makes an angle of less than 45° with the horizontal. (See Figure 2-9.)

Hose Bibb. An end faucet to which a hose may be attached, usually having a $3/4$-inch pipe size with 12 threads per inch. (Also referred to as *Sill Cock*.)

Hot Water. Water at a temperature greater than or equal to 110°F (43°C).

House Drain. See *Building Drain*.

House Sewer. See *Building Sewer*.

House Trap. See *Building Trap*.

Hydrostatic Test. A water test applied using a pumping mechanism to raise pressure above source pressure.

Icemaker Box. A water-outlet box, usually flushed with the exterior wall, designed to serve ice-making equipment or fixtures.

Impervious. A smooth surface, usually watertight.

Indirect Waste Pipe. A waste pipe that does not connect directly with the drainage system, but that discharges into the drainage system through an air break or air gap into a trap, fixture, receptor or interceptor. (See Figure 2-7.)

Individual Sewage Disposal System. A system for disposal of domestic sewage by means of a septic tank, cesspool or mechanical treatment, designed for use apart from a public sewer to serve a single establishment or building.

Individual Vent. A pipe installed to vent a fixture trap that connects with the vent system above the fixture served, or terminates in the open air. (See Figures 2-4 and 4-6.)

Individual Water Supply. A water supply, except an approved public water supply, that serves one or more families.

Interceptor. A device designed and installed to separate and retain for removal, by automatic or manual means, deleterious, hazardous or undesirable matter from normal wastes, while permitting normal sewage or wastes to discharge into the drainage system by gravity.

Invert Elevation. Height reached by the bottom-most measurement of the inside of the pipe.

Isometric Drawing. A three-dimensional drawing showing piping systems, drawn at 30° from the horizontal and true vertical.

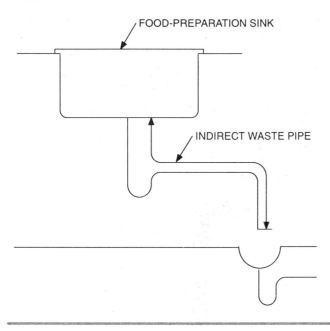

Figure 2-7 *Traps for indirect waste pipes*

Joint

    Expansion. A loop, return bend or return offset that provides for the expansion and contraction in a piping system and is utilized in tall buildings or where there is a rapid change of temperature, as in power plants, steam rooms and similar occupancies.

    Flexible. Any joint between two pipes that permits one pipe to be deflected or moved without movement or deflection of other pipe.

    Mechanical. See *Mechanical Joint.*

    Slip. A type of joint made by means of a washer or a special type of packing compound in which one pipe is slipped into the end of an adjacent pipe.

Knockout Plug. A plug or cap that is wedged or formed in place, capable of being removed when struck by a hammer or other means.

Lateral. Drainage piping to a main sewer that serves a single building.

Lavatory. Generally the bathroom hand sink used for washing, shaving, brushing teeth, etc. There are institutional lavatories for special purposes.

Leader. An exterior drainage pipe for conveying storm water from roof or gutter drains to an approved means of disposal. (See Figure 2-9.)

Lead-free Pipe and Fittings. Containing not more than 8.0 percent lead.

Lead-free Solder and Flux. Containing not more than 0.2 percent lead.

Liquid Petroleum Gas (LPG). Either propane, propylene, butane or other gases in the same family that can be made into liquid form for the purpose of transporting or storage.

Listing Agency. The agency accepted by the state, county or local government authorities that lists and labels products during an inspection and testing procedure and usually follows up with a published report stating the results of those procedures.

Local Vent Stack. A vertical pipe to which connections are made from the fixture side of traps and through which vapor or foul air is removed from the fixture or device used on bedpan washers. (See Figure 2-9.)

Loop Vent. The reventing of a horizontal branch drain to the original vent stack parallel with the waste stack.

Main. The principal pipe artery to which branches are connected. (See Figure 2-9.)

Main Vent. The principal artery of the venting system, to which the vent branches may be connected. (See Figure 2-9.)

Manhole. An access opening through which a person can gain access to a vault or enclosed area.

Manifold. See *Plumbing Appurtenance.*

Mechanical Joint. A connection between pipes, fittings, or pipes and fittings that is not screwed, caulked, threaded, soldered, solvent cemented, brazed or welded. A joint in which compression is applied along the centerline of the pieces being joined. In some applications, the joint is part of a coupling, fitting or adapter.

Medical Gas System. The complete system to convey medical gases for direct patient applications from central supply systems (bulk tanks, manifolds and medical air compressors) with pressure and operating controls, alarm warning systems, and related components, and piping networks extending to station outlet valves at patient use points.

Medical Vacuum Systems. A system consisting of central-vacuum-producing equipment with pressure and operating controls, shutoff valves, alarm-warning systems, gauges and a network of piping extending to and terminating at suitable station inlets at locations where patient suction may be required.

Minor Repairs. Repairs that are noninvasive and that do not require a permit. Drain cleaning, faucet and out-of-wall fixture replacement, small piping repairs and valve-stem replacement are examples.

Mop Sink. A deep-well sink supplied with water that is backflow protected, used for the process of cleaning floors, etc.

National Pipe Thread (NPT). The standard for pipe threading iron pipe as well as standard size for fittings.

Nonpotable Water. Water not safe for drinking, personal or culinary use.

Nuisance. Public nuisance as known in common law or in equity jurisprudence; whatever is dangerous to human life or detrimental to health; whatever structure or premise is not sufficiently ventilated, sewered, drained, cleaned or lighted, with respect to its intended occupancy; and whatever renders the air, human food, drink or water supply unwholesome.

Occupancy. The purpose for which a building or portion thereof is utilized or occupied.

Offset. A combination of approved bends that makes two changes in direction bringing one section of the pipe out of line, but into a line parallel with the other section.

Open Air. Outside the structure.

Orifice. A hole or opening that allows the flow of air or gas, regulated by the size of the hole, during the operation of equipment or appliances.

O-ring. A round sealing washer or gasket used in fixtures, faucets and related ware.

Overflow. An outlet on a tub or sink that allows water or liquid to flow into an auxiliary drain below the rim of the fixture served. (See Figure 2-8.)

Packing. A sealant material usually pressed into a valve-type body to provide a water- or air-tight joint.

Pipe Covering. Insulation protecting the contents of such pipe.

Plumbing. The practice, materials and fixtures utilized in the installation, maintenance, extension and alteration of all piping, fixtures, plumbing

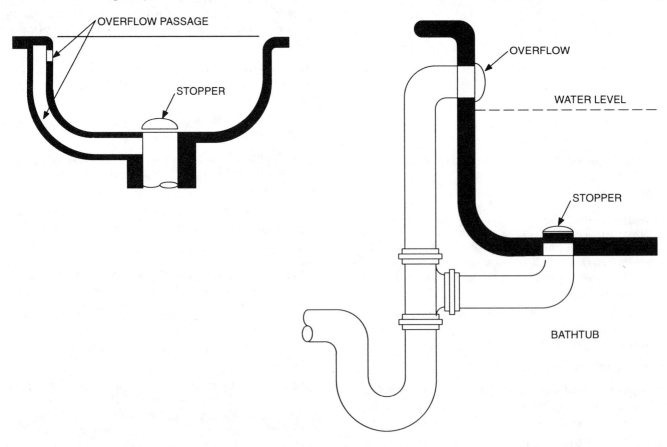

Figure 2-8 *Fixture overflow*

appliances and plumbing appurtenances, within or adjacent to any structure, in connection with sanitary drainage or storm drainage facilities, venting systems, and public or private water supply systems.

Not included in this definition are installations or chilled-water piping in connection with refrigeration, process and comfort cooling; hot water piping in connection with building heating; and piping for fire-protection systems.

Plumbing Appliance. Any one of a class of plumbing fixtures intended to perform a special function. Included are fixtures having the operation or control dependent on one or more energized components, such as motors, controls, heating elements, or pressure- or temperature-sensing elements. Such fixtures are manually adjusted or controlled by the owner or operator, or are operated automatically through one or more of the following actions: a time cycle, a temperature range, a pressure range, a measured volume or weight.

Plumbing Appurtenance. A manufactured device, prefabricated assembly, or an on-the-job assembly of component parts that is an adjunct to the basic piping system and plumbing fixtures. An appurtenance demands no additional supply and does not add any discharge load to a fixture or to the system.

Plumbing Fixture. A receptacle or device that is either permanently or temporarily connected to the water distribution system of the premises and demands a supply of water there from; discharges waste water, liquid-borne waste materials or sewage, either directly or indirectly to the drainage system of the premises; or requires both a water-supply connection and a discharge to the drainage system of the premises.

Plumbing System. Includes the water supply and distribution pipes; plumbing fixtures and traps; water-treating or water-using equipment; soil, waste and vent pipes; and sanitary and storm sewers and building drains, in addition to their respective connections, devices and appurtenances within a structure of premises. (See Figure 2-9.)

Pollution. An impairment of the quality of potable water to a degree that does not create a hazard to the public health but that does adversely and unreasonably affect the aesthetic qualities of such potable waters for domestic use.

Potable Water. Water free from impurities present in amount sufficient to cause disease or harmful physiological effects, and conforming in bacteriological and chemical quality to the requirements of the Public Health Service Drinking Water Standards or the regulations of the public health authority having jurisdiction.

Pressure Gauge. A device used to measure pressure, liquid or gas, to test the integrity of the plumbing system.

Pressure Regulator. A device that controls pressure either by a constant means or by manual variable means.

Private. In the classification of plumbing fixtures, "private" applies to fixtures in residences and apartments, and to fixtures in nonpublic toilet rooms of hotels and motels, and similar installations in buildings where the plumbing fixtures are intended for use by a family or an individual.

Public or Public Utilization. In the classification of plumbing fixtures, "public" applies to fixtures in general toilet rooms of schools, gymnasiums, hotels, airports, bus and railroad stations, public buildings, bars, public comfort stations, office buildings, stadiums, stores, restaurants and other installations where a number of fixtures are installed so that their use is similarly unrestricted.

Public Water Main. A water-supply pipe for public use controlled by public authority.

Quick-closing Valve. A valve or faucet that closes automatically when released manually or that is controlled by a mechanical means for fast-action closing.

Ready Access. That which enables a fixture, appliance or equipment to be directly reached without requiring the removal or movement of any panel, door or similar obstruction, and without the use of a portable ladder, step stool or similar device.

Reduced Pressure Principle Backflow Preventer. A backflow prevention device consisting of two independently acting check valves, internally force-loaded to a normally closed position and separated by an intermediate chamber (or zone) in which there is an automatic relief means of venting to atmosphere, internally loaded to normally open position between two tightly closing shutoff valves and with means for testing for tightness of the checks and opening of relief means.

Registered Design Professional. An architect or engineer registered or licensed to practice professional architecture or engineering as defined by the statutory requirements of the state in which the project is to be constructed.

Relief Valve

Pressure-relief Valve. A pressure-actuated valve held closed by a spring or other means and designed to relieve pressure automatically when the pressure at which the valve is set is reached.

Temperature- and Pressure-relief Valve (T&P). A combination relief valve designed to function as both a temperature-relief and pressure-relief valve.

Contractor's Guide to the Plumbing Code

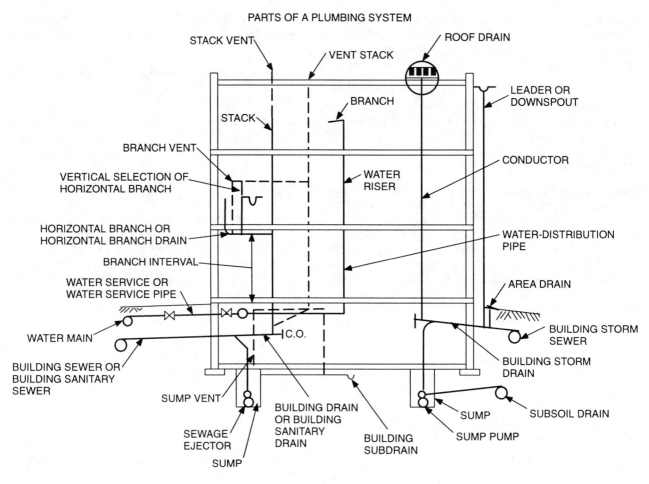

Figure 2-9 *Parts of a plumbing system*

Temperature-relief Valve. A temperature-actuated valve designed to discharge automatically at the temperature at which such valve is set.

Relief Vent. A vent whose primary function is to provide circulation of air between drainage and vent systems. (See Figures 2-4 and 4-10.)

Rim. An unobstructed open edge of a fixture.

Riser. See *Water-pipe Riser*.

Roof Drain. A drain installed to receive water collecting on the surface of a roof and to discharge such water into a leader or a conductor. (See Figure 2-9.)

Rough-in. Parts of the plumbing system that are installed prior to the installation of fixtures. This includes drainage, water supply, vent piping, and the necessary fixture supports and any fixtures that are built into the structure. Also called top out.

Sanitary. A building sewer that conveys sewage only.

Self-closing Faucet. A faucet containing a valve that automatically closes upon deactivation of the opening means.

Separator. See *Interceptor*.

Septic Tank. The receptor that holds the waste water from the building drain. It then allows effluent to exit into leach lines, thus holding solids for their decomposition.

Sewage. Any liquid waste containing animal or vegetable matter in suspension or solution, including liquids containing chemicals in solution.

Sewage Ejectors. A device for lifting sewage by entraining the sewage in a high-velocity jet of steam, air or water. (See Figure 2-9.)

Sewer (See Figure 2-9.)

    Building Sewer. See *Building Sewer*.

Public Sewer. A common sewer directly controlled by public authority.

Sanitary Sewer. A sewer that carries sewage and excludes storm, surface and ground water.

Storm Sewer. A sewer that conveys rainwater, surface water, condensate, cooling water or similar liquid wastes.

Sill Cock. See *Hose Bibb*.

Slope. The fall (pitch) of a line of pipe in reference to a horizontal plane. In drainage, the slope is expressed as the fall in units vertical per units horizontal (percent) for a length of pipe.

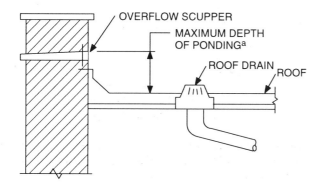

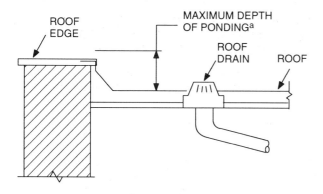

a. The level must rise above the weir to induce flow. The level will continue to rise until the discharge and rainfall rate reach equilibrium. The maximum depth must be considered in calculating roof loads.

Figure 2-10 *Roof drain with overflow*

Soil Pipe. A pipe that conveys sewage containing fecal matter to the building drain or building sewer.

Spill-proof Vacuum Breaker. An assembly consisting of one check valve force-loaded closed and an air-inlet vent valve force-loaded open to atmosphere, positioned downstream of the check valve, and located between and including two tightly closing shutoff valves and a test cock.

Stack. A general term for any vertical line of soil, waste, vent or inside conductor piping that extends through at least one story with or without offsets. (See Figure 2-9.)

Stack Vent. The extension of a soil or waste stack above the highest horizontal drain connected to the stack. (See Figure 2-9.)

Stack Venting. A method of venting a fixture or fixtures through the soil or waste stack.

Standard Pipe Weight. The weight of pipe characterized by the wall thickness, i.e., Schedule 20, 30, 40, 80, etc. Schedule 40 is considered standard weight.

Sterilizer

   Boiling Type. A nonpressure-type fixture utilized for boiling instruments, utensils or other equipment. It is used for disinfecting. These devices are portable or are connected to the plumbing system.

   Instrument. A device for the sterilization of various instruments.

   Pressure (autoclave). A pressure-vessel fixture designed to utilize steam under pressure for sterilizing.

   Pressure Instrument Washer. A pressure-vessel fixture designed to both wash and sterilize instruments during the operating cycle of the fixture.

   Utensil. A device for the sterilization of utensils used in health care services.

   Water. A device for sterilizing water and storing sterile water.

Sterilizer Vent. A separate pipe or stack, indirectly connected to the building drainage system at the lower terminal, that receives the vapors from nonpressure sterilizers, or the exhaust vapors from pressure sterilizers, and conducts the vapor directly to the open air. Also called vapor, steam, atmospheric or exhaust vent.

Storm. A building sewer that conveys storm water or other drainage, but not sewage.

Storm Drain. See *Drainage System, Storm*.

Strapping Tape. Thin-gauge metal straps used to stabilize or hold piping in place.

Structure. That which is built or constructed, or a portion thereof.

Subsoil Drain. A drain that collects subsurface water or seepage water and conveys such water to a place of disposal. (See Figure 2-9.)

Sump. A tank or pit that receives subsurface drainage sewage or liquid waste, located below the normal grade or below the gravity system, and that must be emptied by mechanical means. (See Figure 2-9.)

Sump Pump. An automatic water pump powered by an electric motor for the removal of drainage, except raw sewage, from a sump, pit or low point. (See Figure 2-9.)

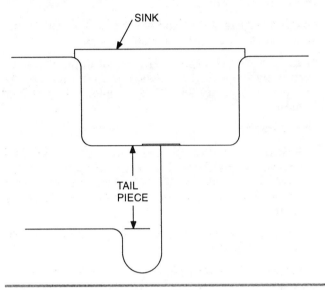

Figure 2-11

Sump Vent. A vent from a pneumatic sewage ejector, or similar equipment, that terminates separately to the open air. (See Figure 2-2.)

Supports. Devices for supporting and securing pipe, fixtures and equipment.

Swimming Pool. Any structure, basin, chamber or tank containing an artificial body of water for swimming, diving or recreational bathing, having a depth of 2 feet or more at any point.

Tailpiece. The drainage tubing between the fixture and the trap served. (See Figure 2-11.)

Tempered Water. Water having a temperature range between 85°F (29°C) to 110°F (43°C).

Third-party Certification Agency. An approved agency operating a product or material certification system that incorporates initial product testing, assessment and surveillance of a manufacturer's quality control system.

Third-party Certified. Certification obtained by the manufacturer indicating that the function and performance characteristics of a product or material have been determined by testing and ongoing surveillance by an approved third-party certification agency. Assertion of certification is in the form of identification in accordance with the requirements of the third-party certification agency.

Third-party Tested. Procedure by which an approved testing laboratory provides documentation that a product, material or system conforms to specified requirements.

Trap. A fitting or device that provides a liquid seal to prevent the emission of sewer gases without materially affecting the flow of sewage or waste water through the trap. (See Figure 2-5.)

Trap Arm. The piping between the trap and the vent for that fixture. Also known as a fixture drain. (See Figure 2-12.)

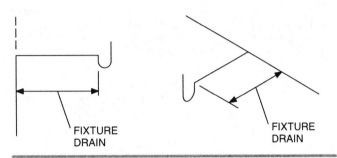

Figure 2-12 *Fixture drain (also known as a trap arm)*

Trap Seal. The vertical distance between the weir and the top of the dip of the trap. Also known as fixture drain. (See Figure 2-5.)

Type A Dwelling Unit. A dwelling unit that is designed and constructed for accessibility in accordance with the provisions of ICC/ANSI A117.1–98.

Type B Dwelling Unit. A dwelling unit that is designed and constructed for accessibility in accordance with the provisions of ICC/ANSI A117.1–98.

Unstable Ground. Earth that does not provide a uniform bearing for the barrel of the sewer pipe between the joints at the bottom of the pipe trench.

Vacuum. Any pressure less than that exerted by the atmosphere.

Vacuum Breaker. A type of backflow preventer installed on openings subject to normal atmospheric pressure that prevents backflow by admitting atmospheric pressure through ports to the discharge side of the device.

Vent Pipe. See *Vent System*.

Vent Stack. A vertical vent pipe installed primarily for the purpose of providing circulation of air to and from any part of the drainage system. (See Figure 2-9.)

Vent System. A pipe or pipes installed to provide a flow of air to or from a drainage system, or to provide a circulation of air within such system to protect trap seals from siphonage and backpressure. (See Figure 2-9.)

Vertical Pipe. Any pipe or fitting that makes an angle of 45° or more with the horizontal.

Wall-hung Water Closet. A wall-mounted water closet installed in such a way that the fixture does not touch the floor.

Waste. The discharge from any fixture, appliance, area or appurtenance that does not contain fecal matter. (See Figure 2-9.)

Waste Pipe. A pipe that conveys only waste. (See Figure 2-9.)

Water Heater. Any heating appliance or equipment that heats potable water and supplies such water to the potable hot-water distribution system.

Water Main. A water-supply pipe or system of pipes, installed and maintained by a city, township, county, public utility company or other public entity, on public property, in the street or in an approved dedicated easement of public or community use. (See Figure 2-9.)

Water Outlet. A discharge opening through which water is supplied to a fixture, into the atmosphere (except into an open tank that is part of the water supply system), to a boiler or heating system, or to any devices or equipment requiring water to operate, but which are not part of the plumbing system.

Water Softener. Tanks using membranes that, through chemical process, reduce the hardness in the water.

Water-distribution Pipe. A pipe within the structure or on the premises that conveys water from the water-service pipe, or from the meter when the meter is at the structure, to the points of use. (See Figure 2-9.)

Water-hammer Arrester. A device used to absorb the pressure surge (water hammer) that occurs when water flow is suddenly stopped in a water supply system. (See Figure 5-3.)

Water-pipe Riser. A water-supply pipe that extends one full story or more to convey water to branches or to a group of fixtures. (See Figure 2-9.)

Water-service Pipe. The pipe from the water main or other source of potable water supply, or from the meter when the meter is at the public right of way, to the water-distribution system of the building served. (See Figure 2-9.)

Water-supply System. The water-service pipe, water-distribution pipes, and the necessary connecting pipes, fittings, control valves, and all appurtenances in or adjacent to the structure or premises.

Well

Bored. A well constructed by boring a hole in the ground with an auger and installing a casing.

Drilled. A well constructed by making a hole in the ground with a drilling machine of any type and installing casing and screen.

Driven. A well constructed by driving a pipe into the ground. The drive pipe is usually fitted with a well point and screen.

Dug. A well constructed by excavating a large diameter shaft and installing a casing.

Wet Vent. A vertical drain that drains into another vertical drain vent on the same floor level.

Whirlpool Bathtub. A plumbing appliance consisting of a bathtub fixture that is equipped and fitted with a circulating piping system designed to accept, circulate and discharge bathtub water upon each use.

Yoke Vent. A pipe connecting upward from a soil or waste stack to a vent stack for the purpose of preventing pressure changes in the stacks.

3

Documentation, Standards, Alternatives and Inspections

In this chapter, we continue the discussion on getting a permit, what documentation you need and what the code department may need from you in order to prove that the products or practices you propose are acceptable. The types of inspections that will be required are also discussed.

General

Now it's almost time to start the installation. In this section we're going to examine what will be required prior to starting. We will discuss planning for the installation, and this is where you determine the types of materials that you will use. Of course, in terms of penetration protection, this will be subject to the type of construction and occupancy or use the architect or designer chooses. The intended use of the plumbing system will also affect the materials chosen.

We'll examine how plans and schematics can help out in this area. We'll also discuss legal requirements, including obtaining the permit, as we mentioned in Chapter 1. This will involve purchasing the permit, determining if there is any licensing that may be involved, as well as any state or local amendments.

The resources that are available to assist you will also be examined—mainly the code official, or his or her assistants, that we talked about previously. An appeal process may also apply should you and the code official disagree on any specific subject (except administrative rules).

Contractor's Guide to the Plumbing Code

There is also a process available whereby you can submit proposed alternative methods or materials for consideration and approval. These may or may not be expressly covered in the code, but if you can provide sufficient tests or evaluation reports to prove that they meet the intent of the code, then the code official can review them prior to approval or denial.

Finally, we will discuss things to remember as you begin the actual installation, and other little tips that will help you carry out the installation in a straightforward manner while incurring the fewest problems.

Getting the Permit

Let's first examine what is required prior to starting the installation. This is what we might refer to as the planning stage. Recall that in the first chapter we discussed the need for a permit. Section 106.3 explains how to apply for that permit. It says, in part:

> "Each application for a permit, with the required fee, shall be filed with the code official on a form furnished for that purpose and shall contain a general description of the proposed work and its location."

Most of these forms will be pretty much self-explanatory. Nonetheless, the code official, or his or her staff, will be available to assist you in filling out "that form." It may be helpful to get a copy of the application ahead of time in order to review it carefully prior to submittal. While many jurisdictions allow applications for the plumbing permit to be purchased over the counter without a plan review, giving wrong information or a lack of required information can result in a frustrating and wasted trip. Just as with the actual installation, hurrying to accomplish the task without doing sufficient homework or preparation, including your initial paperwork, can cause needless delays. You may end up directing your frustration at the code official, but this may hinder future valuable communication that you really want and need.

If you have already purchased the building permit for the construction and you are merely applying for another permit to install the plumbing, you would do well to have a copy of the original building permit when filling out the application. This building permit will likely have the answers to any questions that may arise in the process. Remember, the better prepared you are, the smoother the process will be.

Because no specific fee table is given in the IPC, the cost of the permit may vary from jurisdiction to jurisdiction. In some cases the fee will be based on a cost per fixture with an accompanying base permit fee. Of course, if you're dealing with a commercial type of building that may have various occupancies located within one building, you will need to identify these types of occupancies clearly on any plans or permit application for the benefit of the code official during review.

Documentation

This brings us to the matter of construction documents that may be required with the plumbing permit. Again, this will vary from jurisdiction to jurisdiction, so check locally. These documents, as explained in Section 106.3.1, may include engineering calculations, line diagrams, specifications or other such data. You will need to submit two or more sets of this material with your application for the permit. Whether required or not, such plans can be very helpful in designing the system. In Chapter 4 we will examine how you can draw a schematic to assist you during installation. A typical schematic (see Figures 3-1A and 3-1B) can be helpful in determining pipe sizing, types of fittings, location of cleanouts, etc. Why not ask the code official to review your schematic with you? It could save you time and money. Major problems can be resolved at this stage by noting any noncompliant portions of the installation. Then you can find out how to do it properly before you install it.

Remember, do not install the plumbing prior to permit issuance. It will be costly both in penalties and delays, and may also require tearing out incorrectly installed systems.

When a schematic is required along with the application and you have already installed the system, don't simply draw an existing noncompliant installation just because you did the work already. This can prove to be very frustrating for everyone involved!

As the systems get more complicated, more detailed plans are required. You will need to give attention to such things as the type of fixtures to be installed, the type of construction and what materials would be suitable, the location of necessary valves, safety devices, etc., and the protection of penetrations through fire-resistive construction, not to mention sizing of drains, vents, water lines, and other details.

If your installation involves of a large quantity of fixtures or numerous types of occupancy classifications, you may want to consult a professional designer to prepare such plans and calculations. In other words, if you haven't been able to figure it out by reading the IPC, consulting the code official, or using reference guides such as this one, it's probably over your head! You don't want to have the code official inspect the installation and then receive a correction note that reads "I suggest you get a plumber!" Get some help and save yourself time, money and frustration. Contrary to popular opinion, plumbing knowledge is not limited to "waste runs downhill."

There is also the matter of licensing and certification requirements. You need to find out what state or local regulations may apply. This would involve not only requirements for business or contractor licensing and construction bonds, but also any license or certification that may be required to physically install the plumbing. Many states require a journeyman level certificate. There may be exemptions for property owners or their agents or laborers doing work outside the building up to the property line. Ask the code official if you are not sure.

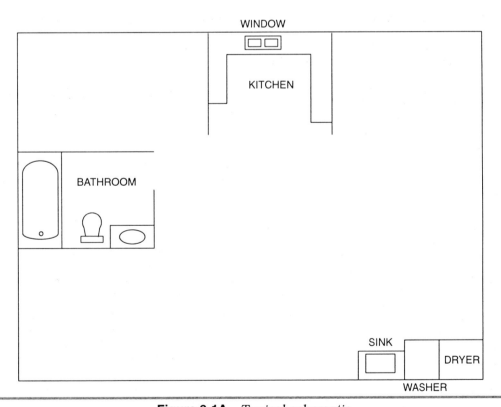

Figure 3-1A *Typical schematic*

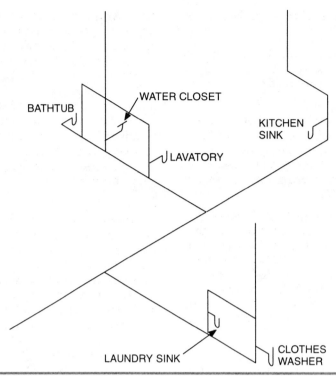

Figure 3-1B *Typical schematic*

The Code Official

Let's now talk more about the code official. We briefly touched on some responsibilities of the code official, but let's examine in detail how this official can be a valuable resource in your endeavor. While a plumber has a good understanding of how the plumbing system operates, the code official and staff have much more knowledge of the overall construction of the site and the structure, and the interrelationship between it and the other life-safety provisions of the zoning, building, mechanical and fire codes.

There is more to plumbing than just pipes and fittings installed per the "guidelines" of a plumbing code. The construction of a building involves excavation, structural details, and mechanical systems for heating, cooling, and ventilation, just to name a few. There are additional codes that regulate these aspects of construction, and the code official has to understand how all of these codes operate in unison. Needless to say, a family of codes, such as those offered by the International Code Council and the publisher of the IPC, coordinated with each other to eliminate conflicts, is a must for building administration.

The code department staff is going to be interested in the types of materials you are planning to install. This is a subject that requires close attention. There are few things worse than completing the installation only to learn that the wrong materials were used and everything must be torn out and replaced. If you felt uncomfortable doing the plumbing up to that point, you may just give up. Please be aware that the materials you choose must be *approved* for such use.

Standards and Approvals

Well, what does "approved" mean? In order to understand, we need to know about "standards."

"Approved" refers to the material or product having been tested for "quality, strength, effectiveness, fire-resistance, durability and safety" as prescribed in Section 105.2. Chapter 13 lists standards for various plumbing materials. Generally, these standards are rules and commitments made between manufacturers, suppliers and plumbing installers to ensure a quality product. However, *quality control* is essential to get this assurance. As President Ronald Reagan used to quote from an old Russian saying, "Trust, but verify!"

Let's discuss an experience one of the authors of this publication had that will help you understand standards. Notice in Chapter 13, the name of the agency, ANSI, which stands for The American National Standards Institute. Now find the standard number for plastic shower receptors and shower stalls. The number given is Z124.2 and it refers you to Section 417.1. There it reads in part:

"Prefabricated showers and shower compartments shall conform to ANSI Z124.2, ASME A112.19.9 or CSA B45.5."

So, the fiberglass shower unit that you wish to install should be listed by one of these three standards agencies. But you've already purchased and installed the unit! Now what? That is exactly what happened in this example. When the inspector could not locate a listing, he could not approve the installation. The first reaction might be to rip it out and get a shower unit that is tested and approved. But let's back up a minute and see if there is another way to resolve this problem.

Previously, in our discussion of the meaning behind "approved," we referred to a product having to be tested for quality, strength, durability, and so on. Well, that is where the code official has the discretionary authority to accept alternate materials and methods as outlined in Section 105. The author relied on the criteria in Section 105.3 that allows the code official to:

"require tests as evidence of compliance to be made at no expense to the jurisdiction."

The first thoughts that occur include *How would I test it? Jump up and down in it? Thrust my body against the enclosure walls?* This may have been a good test, but if the shower unit failed, it would be difficult to defend the testing. So, having contacted the manufacturer of the shower enclosure, the code offical informed him that an identically manufactured unit must be sent to a testing laboratory to see if it would pass the required test, and he must forward the results. What was received was a document from an independent testing lab outlining the test procedure of, you guessed it, ANSI Z124.2. And guess what it involved? The procedure required weighted sandbags to fall on the unit as well as suspended sandbags to be swung to hit the walls at various angles. Not much different from what one could have done with his or her body.

There were also heat tests, leak tests, even a "cigarette burn" test. (Who ever would have thought of that one!) The bottom line is that the unit proved to be equivalent to the standards in quality, strength, effectiveness, fire-resistance, durability and safety. A word of caution though—make sure that the material or method has been tested and listed for the relevant and specific purpose intended. For instance, a copper fitting may have an approved listing for drainage, waste and vent installation but probably would not be approved for potable water systems.

What recourse existed had the code official not chosen to accept a test for equivalency? Here is where Section 109 comes into play. You have the right to appeal. The appeals board is made up of qualified individuals representing various facets of construction trades. Sections 109.3 through 109.7 define the operations and representation of the board of appeals, but this may vary according to jurisdictional authority. If you wish to go this route, contact the code official who can explain these aspects to you.

Since this is an administrative process, there may be a modest fee associated with an application for appeal. The provision for appeal does not mean, however, that any required correction needed, as identified from a result of plan review or inspection, is a proper matter for appeal. As stated in the IPC Commentary (ICC, 1998), "The intent of the appeal process is not to waive or set aside a code requirement; rather, it is intended to provide a means of reviewing a code official's decision on an interpretation or application of the code or to review the equivalency of protection to the code requirements." So the appeal process provides a forum outside of a court's jurisdiction in which to review the code official's determination.

On the other hand, if you feel, for instance, that it is unfair that the code official requires you to test your piping installation when you have never had a leak, you will not likely be granted an appeal. The appeal process does not provide a way of "relaxing" the code provision. A jurisdictional board of appeals, patterned after the provisions of Section 109 of the IPC, provides a workable process of appeal because the board has the authority to modify or reverse the decision of the code official where it is reasonable and legal to do so. And, remember, you still have the right to appeal a decision of the board to the local court system.

Materials, Plans and Preliminary Layout

So, now that you have done your homework away from the job site, let's discuss what you need to do at the site in preparation for the plumbing installation.

One of the first things you will need to determine is the type of material you will use. This is mostly dependent on the type of construction, intended use and costs. (See Chapters 4, 5 and 6 for details.)

Then you will need to locate the sewer or plan for a septic system and determine the source of water, be it from a water purveyor or from a well, cistern or tank. Once you have located these items you can begin drawing the schematic for the plumbing installation. Keep in mind however, that the depth of the sewer/septic hook-up must be ascertained. With this information you can establish the location and depth of the building drain stub-out for the building. In Chapter 4 we will discuss this further.

Another factor in determining the layout of the plumbing installation is the framing of the walls and floors. The Building Code provides criteria for notching, cutting and boring of framing members. Figures 3-2 through 3-5 show some of the common examples of minimum and maximum allowances in accordance with the governing building code. This information is also found in Appendix F of the IPC.

In a nutshell, if you are going to run horizontal piping of $1^1/_2$-inch diameter through a bearing stud wall, a 2-inch by 4-inch stud wall will be inadequate. If you are constructing the building yourself you can make sure you accommodate the plumbing. If not, contact the carpenters and make them aware of your intention to somewhat deplete the integrity of the structural system. This is because they did not provide adequately sized framing for the pipe penetrations. (Truth to tell, it must be admitted that there is a definite euphoric buzz a plumber gets from sawing through a substantial piece of wood!)

This all emphasizes the need to be familiar with the approved plans for the building's structure. Possible roadblocks can be identified prior to installation if you take the necessary time to become familiar with the approved set of plans. And do not take it for granted that any set of plans for the structure will do. You need to look at the plans approved and stamped by the code official! These plans will indicate items that might or might not otherwise be shown on additional plans. Imagine the cost and time delay you would have if the approved plans indicated a particular fire-resistive construction of a wall that was not shown to have any fire rating on the plans you looked at. It is in your best interest to ask for the approved set of plans in your preparation.

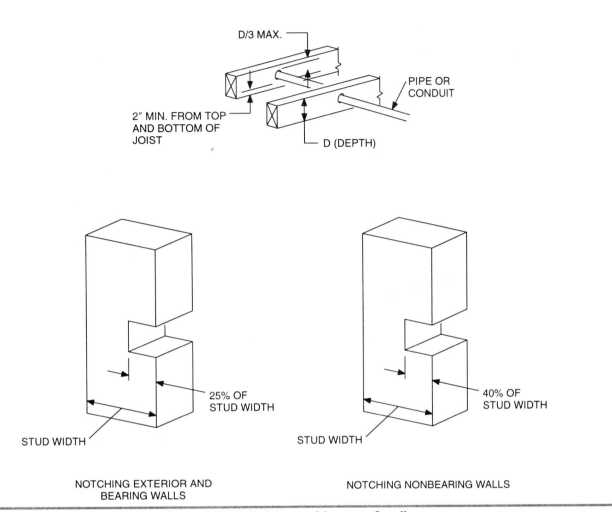

Figure 3-2 *Notching and boring of walls*

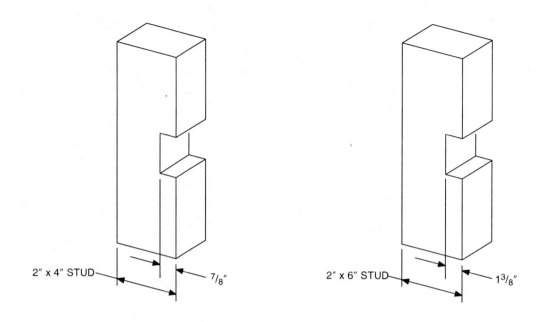

NOTCHING EXTERIOR AND BEARING WALLS

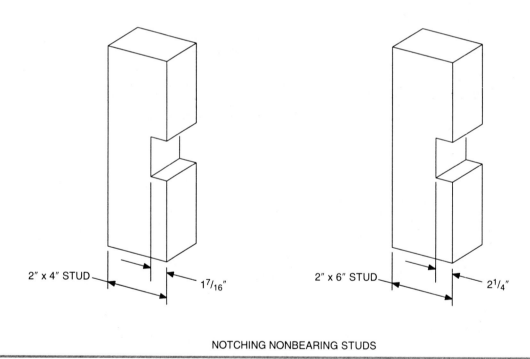

NOTCHING NONBEARING STUDS

Figure 3-3 *Notching and boring of studs and walls*

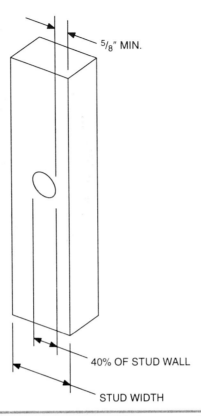

Figure 3-4 *Boring any bearing stud wall*

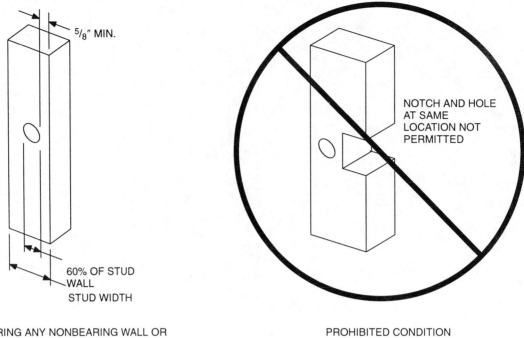
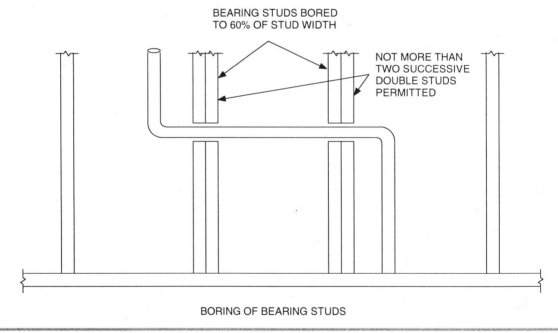

Figure 3-5 *Boring of studs and walls*

Documentation, Standards, Alternatives and Inspections

Required Inspections

A subject you should become familiar with prior to installation is required inspections. Section 107.1 outlines the minimum types of inspections that you should expect during the installation procedure:

- The first inspection will be the **underground** inspection. This is done after all the piping is installed in the trenches and any backfill is put into place. Before the installation is covered, and prior to the inspection, you will need to test the pipe and fittings. Typically this is done by capping the lowest end of the pipe (ideally at the point of connection to the side sewer outside the building) and filling the system with water to a height of 10 feet above the highest joint. An air test of 5 pounds per square inch (psi) can be applied [except for polyvinyl chloride drain, waste and vent (PVC DWV)]. This may prove handy where water is unavailable, or in an area subject to freezing. A word of caution though: If you test with air, keep in mind that certain piping systems filled with air are likely to float if the trenches become filled with water. Even if the piping is filled with water and there is water in the trench, the piping may still dislodge. Be sure to secure the installation in place.

- The second inspection will be the **rough-in inspection**. This should occur after all piping that will be concealed within the construction of any wall, ceiling, or floor is completed. A water test can be performed by filling the system to the highest point, most likely at the vent penetration through the roof. If the "rough-in" is checked in sections, such as in multi-story construction, then the 10-foot head of water prescribed for testing underground installations shall apply. Here again an air test of 5 psi is an alternative, with the exception of PVC DWV.

- The **final inspection** is performed after all the plumbing, including the fixtures, is completed and the structure is ready to occupy. Most often this will involve a visual test of the plumbing system to verify that the fixtures and their supply piping and drainage piping operate as designed.

During these three phases of inspection you will also be required to test the potable water system. The system may be tested at the working pressure provided from the water purveyor, or an air test of 50 psi can be substituted, except for plastic piping. Check with the code official for test pressure required where piping is brazed and will be buried beneath a concrete floor. Also, be aware that some plastic pipe manufacturers require a hydrostatic test using water for joints beneath a concrete slab, in some cases as much as 150 psi.

Aside from these necessary inspections, you will want to confer with the code official to determine if there are any other inspections required by the jurisdiction, such as a sub-floor inspection or certain safety/health inspections.

Now that we have done our homework in planning our project, let's go back to the recommendation for drawing a schematic. We'll prepare this schematic as we begin detailing the drain, waste and vent system in Chapter 4.

Reference:

1. *1997 International Plumbing Code Commentary.* Falls Church, VA. International Code Council. 1998 (pg.18)

4

Conventional Drain, Waste and Vent Systems

This chapter will explain how to apply the conventional rules for venting, traps and waste disposal, including indirect or special waste disposal systems in the IPC. Chapter 7 of the IPC regulates the installation of sanitary drainage systems. The system design requirements are based on the studies conducted at the National Bureau of Standards, with most of the work spearheaded by Dr. Roy Hunter. More recent research resulted in the allowance of alternative engineered systems. There are supplemental requirements for sanitary drainage systems found in IPC Chapters 8 and 10. Chapter 8 regulates indirect waste systems. Indirect waste is required when there is a need to provide additional safeguards to protect the fixtures from a backup in the sanitary drainage system. Chapter 8 also contains requirements for the installation of special waste and chemical waste systems. Chapter 10 regulates traps and interceptors. Venting provisions are addressed in Chapter 9.

Although these items are discussed in the above-named chapters, we will first bring them all together during this portion of our discussion. You need to understand how these different provisions cooperate to create the plumbing system. Let's examine briefly how they work together. As we do, be sure to refer to the appropriate tables in the IPC as indicated.

We will examine what might be termed a "conventional plumbing system" using plastic pipe and fittings. This type of installation is quite common and is probably the simplest to understand. As we discuss this type of system, we will see how the schematic can help you design the system. Once we have completed the design, we will talk about installation techniques. Then we will explore other possible approaches to venting. You will find that these optional venting provisions offer a varied path to achieving an adequately vented system, which may result in a cost savings along with ease of installation when used with certain types of construction.

Typical Schematic

Please refer now to Figure 4-1. This is a basic schematic of a plumbing system, or the portion of the system known as the drain, waste and vent (DWV). A schematic is really not that difficult to draw. All you need is a drafting triangle (preferably a 30/60/90-degree) and a sharp pencil, and to know in which direction the sewer connection is going to be located.

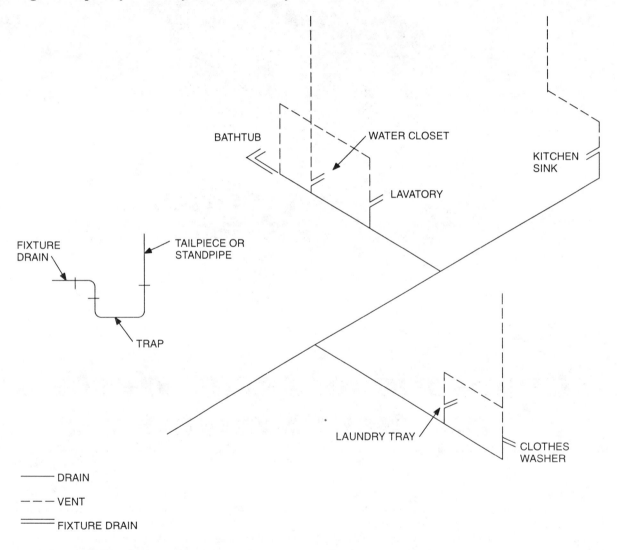

Figure 4-1 *Basic schematic of a plumbing system*

This schematic is taken from the house plan shown in Figure 3-1. Now all you have to remember is that drains must go in the direction of the sewer hook-up and vents must head for the roof, understanding, of course, that drains can connect to other drains and vents to other vents, and that some designed venting systems may not pierce the roof.

You will see that every fixture is going to be provided with a trap. It is called a trap because it traps waste fluids in the waste outlet to create pressure against sewer gases in the drainage system so that they are prevented from coming into the building. Remember, no matter how beautifully you designed the building, if the place smells like a sewer, you are not going to stay there to enjoy the atmosphere, let alone entertain guests! A few fixtures, the common example of which is the toilet, have a built-in trap. But more often than not you will be installing the trap.

Second, the trap (or fixture with a built-in trap) will connect to a fixture drain, sometimes referred to as a trap arm, which will then connect to a vent. The vent is the means by which the trap is protected from a vacuum forming, thus siphoning the fluids from the trap and allowing unwanted gases to escape. If you've ever used a hose to siphon liquid from one container into another, you know that the siphoning action could only occur if the container being siphoned was higher than the other container and the hose did not come out of the liquid and begin drawing air. Just as air in the hose breaks the siphon, the vent introduces air into the fixture drain to prevent siphoning. The fixture drain is regulated in length for the same reason (see Table 906.1). The fixture drain must be graded so that fluids will drain properly. If the horizontal distance between the trap and the vent is too long, the lowest point of the trap weir will be higher than the point of the vent connection (see Figure 4-2).

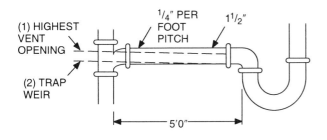

THE TRAP WEIR (2) MUST BE PLACED BELOW THE HIGHEST OPENING TO THE VENT (1)

Figure 4-2 *Distance from trap to vent*

Once the fixture drain is connected to the vent, the piping from that point and downstream is considered the drain. Drains travel by gravity and can be connected with other drains. Vents travel upward toward the atmosphere and can connect to other vents. The fixture drain connects the trap to the drain at the point where the vent is intersected. Let's discuss how the drains can connect with one another.

Drainline Connections

There are two primary areas we must consider when connecting drain lines together. They are (1) the type of fitting to be used at the point of connection, and (2) sizing of the drains.

First of all, if the fittings are not the correct ones for the type of installation, or if they are not installed correctly, the drainage system will fail to work properly. Having to keep a plunger handy can be frustrating and messy, and we all know that calling a plumber to service the system can quickly erode grandma's inheritance. So how do we know which fittings are the right ones for the job?

To answer that question, let's review Table 706.3. As you study the table, compare the plastic fittings shown in Figure 4-3. The table identifies appropriate fittings for three types of connections or changes in direction of drains: (1) horizontal to vertical, (2) vertical to horizontal, and (3) horizontal to horizontal. The fittings listed as bends are just that: they create a bend in the drain line from one direction to another. They are identified by the degree of bend:

- A sixteenth bend being $22^1/_2$ degrees of turn
- An eighth bend being 45 degrees
- A sixth bend being 60 degrees
- A quarter bend being 90 degrees

The short sweep and long sweep fittings are also quarter bends; the terms refer to the short or long radius of turn. (See Table 706.3 on page 80.)

Footnotes "a" and "b" of Table 706.3 provide guidelines for the use of these quarter bends. You should note that the quarter-bend vent fitting shown in Figure 4-3 is not really listed in the table. It is used only on vent piping that is at least 6 inches above the flood level of the fixture being served by the vent. To clear up any confusion about why there are three types of quarter bends listed in the table while only the short and long sweep plastic quarter bends are allowed in drainage systems, the 1997 IPC Commentary states: "Only cast-iron pipe fittings have a separate quarter bend fitting and short sweep fitting. For the other drainage piping materials, a quarter bend and short sweep are the same fitting."

Contractor's Guide to the Plumbing Code

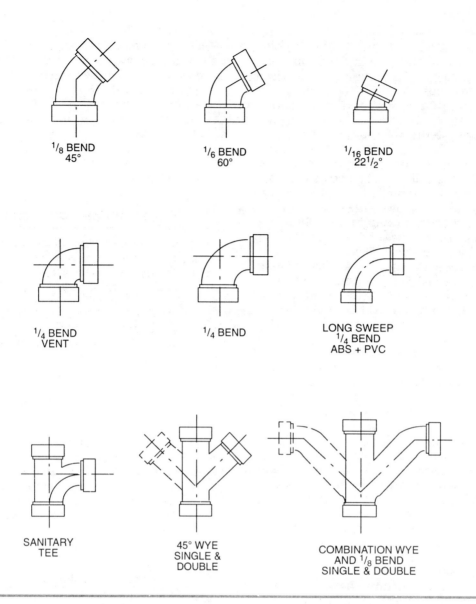

Figure 4-3 *Examples of available fittings*

The last three fittings listed in the table are three-way fittings used to connect three sections of pipe. They can also be found in a double-pattern, or cross, to connect four sections of piping. In such cases, the degree of bend or sweep will be identical on both sides.

Looking at these fittings you can readily see that they need to be installed so that the sweep is in the direction of the drain. In other words, if you want your sewage to end up at the sewage disposal system, point it in that direction. If you point it in the wrong direction, it will either go to the wrong location or it will stay where it is and keep you company while you wait anxiously for the plumber to lay claim on more of your grandma's money. Please take note that the sanitary tee allows only one change of direction in the drainage system, and that is how it is depicted in Figure 4-3, in the vertical position.

Now, we can see the practical application of the fittings in Figure 4-4. The fittings for the vent piping are not listed, but then any of the drainage fittings may be used. As long as the vent fittings are at least 6 inches above the flood level of the fixture served, sweep type of fittings are not necessary. Note also that a long sweep fitting is required on the vertical to horizontal connection of the kitchen sink drain pipe, whereas a short sweep quarter bend is allowed at the water closet and bathtub. This is because the short sweep fitting at the bathtub conforms to footnote "a" of Table 706.3. That is, it is used on a fixture drain and is smaller than 2 inches in diameter. The kitchen sink drain is a drain pipe, not a fixture drain, and thus is not exempt from the long sweep requirement. The short sweep quarter bend at the water closet conforms to footnote "b" of Table 706.3, being 3 inches in diameter and changing from vertical to horizontal in direction.

Conventional Drain, Waste and Vent Systems

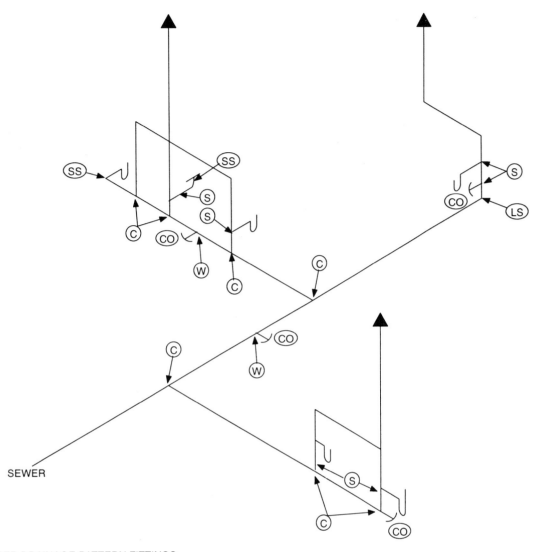

PROPER DRAINAGE PATTERN FITTINGS—
TABLE 706.3

- C = COMBINATION WYE & 1/8 BEND
- W = WYE
- S = SANITARY TEE
- SS = SHORT SWEEP QUARTERBEND
- LS = LONG SWEEP QUARTERBEND
- CO = CLEANOUT

Figure 4-4 *Proper drainage pattern fittings*

Sizing the Drainage System

Are we having fun or what? Now let's talk about sizing the drain lines and then we'll discuss the location and sizing of cleanouts. We will be using Tables 709.1 and 710.1(1), along with Figure 4-5.

Table 709.1 provides fixture-unit values (FUV) for individual fixtures. In order to size the drain lines we need to determine the fixture units that are being drained in each section of pipe. According to Table 709.1, the kitchen sink and dishwasher combined equals 2 units, the laundry sink is 2 units, the clothes washer is 2 units, the bathtub is 2 units, the water closet is 3 units, and the lavatory is 1 unit, for a total of 12 units.

Note that the three bathroom fixtures can be combined as a *bathroom group* and would total 6 fixture units. For the purposes of this exercise, though, it really doesn't make a difference, but it is good to review the definition of a bathroom group. It plays an important part in sizing larger systems and in certain venting methods, as we will soon see.

Contractor's Guide to the Plumbing Code

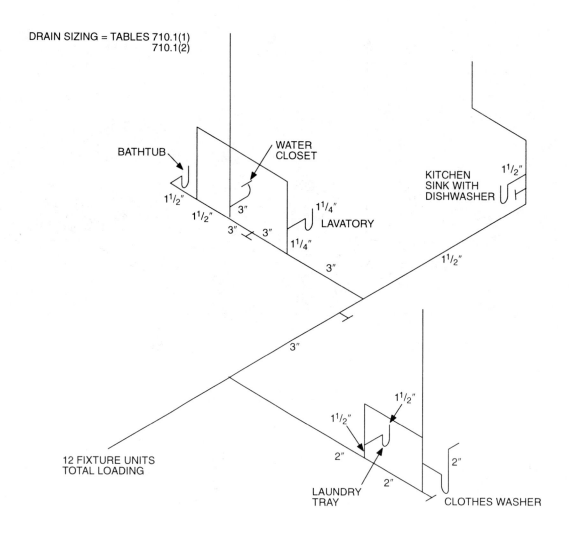

TABLE 709.1 (Fixture Unit Values, FUV)

	FUV	TRAP SIZE
WATER CLOSET	3*	3"
BATHTUB	2*	1 1/2"
LAVATORY	1*	1 1/4"
KITCHEN SINK W/ DISHWASHER	2	1 1/2"
CLOTHES WASHER	2	2"
LAUNDRY TRAY	2	1 1/2"

*For the purposes of sizing the waste branch serving the bathroom, the combined fixtures can be considered a bathroom group at 6 FUV. Each fixture drain and trap must still be sized individually.

Figure 4-5 *Drain sizing*

Let's review Table 710.1(1). This table tells us how many fixture units can be drained into certain sizes of horizontal building drain lines, depending on the slope, or grade, of the piping. Keep in mind that the building drain is the *lowest portion of the horizontal drainage that flows by gravity to the sewage disposal system.* Table 710.1(1) is used to size this portion of the system and all horizontal branches that connect to it. We will plan on grading our installation in this exercise at $1/4$-inch-per-foot grade. (Footnote "a" of this table requires a minimum 3-inch building drain wherever a water closet is served by the system.) Looking at the table, you see that we can load up to 42 fixture units on a 3-inch building drain. That makes it easy because we now have determined that we need a minimum 3-inch drain from the water closet to the sewer connection outside the building. Because we have only 12 fixture units on the system, the 3-inch drain is the largest size we will need to work with. If we had more than 42 fixture units on the system, we would be required to (1) slope the pipe at a $1/2$-inch-per-foot grade that allows up to a maximum load of 50 fixture units, or (2) increase the pipe size, depending on total fixture unit count.

Now let's review all the tables that are used to design the drainage system. Table 709.1 provides the FUVs for the fixtures being drained and tells us the minimum size of fixture trap required for each. Where a fixture is not listed, Table 709.2 is used to calculate the FUVs based on the fixture drain or trap size. Table 906.1 establishes the distance between the trap and its vent. Table 710.1(1) sizes the building drain and any branches of the building drain. This works in conjunction with Table 704.1, which dictates the minimum slope allowed for each diameter of piping. Keep in mind that a horizontal branch may also include a vertical section, such as the vertical sections of the fixture drain piping for those fixtures shown in Figure 4-5. All of these would be sized according to Table 710.1(1). If a vertical drain were to extend to an upper floor it would then qualify as a *stack* (see Chapter 2) and would be sized per Table 710.1(2) along with branches for upper floors and connections to the stack. We will discuss the venting system a bit later.

Cleanouts

One always hopes that, because of a reportedly "superb installation," the drains will never plug up. However, we cannot always control what items may be discharged into the system. Hence, we may have no other alternative then to call a service company to snake out the lines should they become plugged. "Snaking out the lines" is, of course, just a phrase referencing drain-cleaning equipment used in rodding out the lines.

Note that the plumber's fee will likely increase substantially if (1) you stand around and watch him work (just joking), and (2) you do not provide accessibly located cleanouts. Cleanouts are openings in the drainage system appropriately located for the convenience of using the drain cleaning device to dislodge blockages.

Cleanouts (see Section 708) are required every 100 feet on straight horizontal runs, and we are assuming that the horizontal drain lines in Figure 4-5 are less than 100 feet in length. The size of the cleanout is based on the diameter of the drain, but with a 4-inch cleanout being the largest size required. Note that a fixture trap may be disassembled to serve as an approved access for drain cleaning. Drain cleaning equipment is designed to pass through fittings of 45° or less, even if multiple fittings are connected together. Therefore, a cleanout is required wherever a horizontal drain line changes direction more than 45° by the use of one fitting. Where a horizontal drain line changes direction by fittings of 45° or less, a cleanout is required only every 40 feet. Treat the cleanout fittings as you do the drainage fittings we discussed earlier. They need to be installed in relation to the direction of flow within the drain line so that drain cleaning "snakes" will be directed appropriately.

Venting

Now that we have designed our drainage system, there is another ingredient that is absolutely necessary for the plumbing to operate effectively—air! This portion of the plumbing system is called venting. Venting of the sanitary drainage system is one of the most misunderstood areas of plumbing. Venting was invented to protect the trap seal. The odors emanating from the drainage system were kept out of the building by use of a water seal trap. The trap, however, would lose its seal. When venting was added to the drainage system, the trap seal remained in place. The venting requirements in the IPC are designed to maintain the trap seal, and thus prevent the escape of sewer gas. Many studies of venting systems have been made. Chapter 9 of the IPC attempts to recognize all the viable means of venting the drainage system.

There are a number of approaches that can be used in venting a plumbing system. But to begin with, let's consider the easiest—or what might be considered a *conventional venting system*. Table 906.1 provides maximum distance allowable for a vent in relation to the trap. Figure 4-2 shows how and why these distances are necessary. When a fixture vent is connected in this manner, it is called an individual vent. These individual vents are to be sized according to Section 916.2. There we find that the vent must be at least one-half the diameter of the drain served, but in no case less than $1^1/_4$ inches in diameter. That makes things easy!

Remember, the vent size is not based on the trap or fixture drain, but the drain size that continues downstream from the point where the vent and fixture drain connect. The drain size is determined by the fixture-unit load in accordance with Tables 709.1, 709.2, 710.1(1) and 710.1(2). Thus, if the drain size is 4 inches, a 2-inch vent is required. A 3-inch drain would require a $1^1/_2$-inch vent. A 2-inch, $1^1/_2$-inch or $1^1/_4$-inch drain would require a minimum $1^1/_4$-inch vent, this being the minimum size vent allowed.

These individual vents are allowed to connect with each other, but no vent shall connect with another vent unless it is at least 6 inches above the flood rim of the fixture served. In fact, vents must rise vertically to that point. And fittings for vent piping less than 6 inches below the flood rim are to be drainage pattern fittings conforming to Table 706.3. All horizontal vents must grade back to the drain. Sizing of the vents as they connect is again based only on the *required* size of the drain being served (see Figure 4-6). An exception to this is where the developed length of the vent exceeds 40 feet (see Figure 4-7). Looking at Figure 4-8, you can see how these individual vents are sized.

The vertical vents in Figure 4-8, which have additional fixture vents connecting to them, are called *vent stacks* (not to be confused with *stack vents*). The easiest way to remember the difference is to key in to the first word of each phrase. A *vent stack* is a vertical vent in which other vents may connect, whereas a *stack vent* is just that—a vent for a stack. Both of these are sized per Table 916.1. There we can see that the size is based on the developed length of the vent and the number of fixture units being served. Notice in Figure 4-8 that the drain for the clothes washer is 2 inches in diameter. Looking at Table 916.1 we see that a 2-inch drain can have a $1^1/_4$-inch vent stack with a total of 12 fixture units being served. The clothes washer and laundry sink add up to a total of 4 fixture units, so we meet the requirements of the sizing table. If the developed length exceeded 30 feet we would need to increase the diameter of the vent. Section 916.1 states that the minimum size of these vents shall be one-half the diameter of the drain served but in no case less than $1^1/_4$ inches in diameter.

Here is a point to remember when sizing a conventional venting system composed of individual fixture vents, vent stacks and stack vents. The minimum size of an individual vent is one-half the *required drain size*, whereas the minimum size of vent stacks and stack vents is one-half the *size of the drain served*. To see how this is applied, notice the water closets in Figure 4-6, one being drained by a 4-inch-diameter drain and the other by a 3-inch-diameter drain. Both require a minimum individual vent of $1^1/_2$-inch diameter. This is because individual vents are <u>required</u> to be one-half the diameter of the *required drain size*. A water closet requires a minimum 3-inch drain. Now, refer to Figure 4-8 and notice that the vertical vent from the water closet to the point of termination is a vent stack. It must be a minimum of one-half the *diameter of the drain served*. In this case, if the drain for the water closet were 4 inches in diameter, the minimum size of the vent stack would be 2 inches in diameter—half the size of a 4-inch-diameter drain.

Vent Termination

Except where an air-admittance valve is used, which we will address later, vents must terminate to the atmosphere. Just as you don't want to have your home smelling like a sewer on the inside, you most likely would prefer to prevent such wonderful odors from hanging around your yard. Your neighbors would also appreciate that. So, unless you have a death wish (or you want to get back at your neighbor for some reason), you will want to give attention to the distances required for vent terminations.

Vents generally will extend through the roof except where a side wall termination is desired. If the side-wall termination is used, it must be at least 10 feet above ground level. This type of termination must have screens installed to prevent little critters from turning it into their humble abode.

All vent terminations shall be a minimum of 10 feet measured horizontally from any opening into the building, such as a door, window or air intake, unless they are at least 2 feet above the highest portion of such openings. If the roof where a vent extends through for termination is used for any purpose other than weather protection, the vent must terminate at least 7 feet above the surface of the roof. Plus, if you want to use it as an antenna support, better talk to your code official first. Such pipes are of dubious structural strength.

Conventional Drain, Waste and Vent Systems

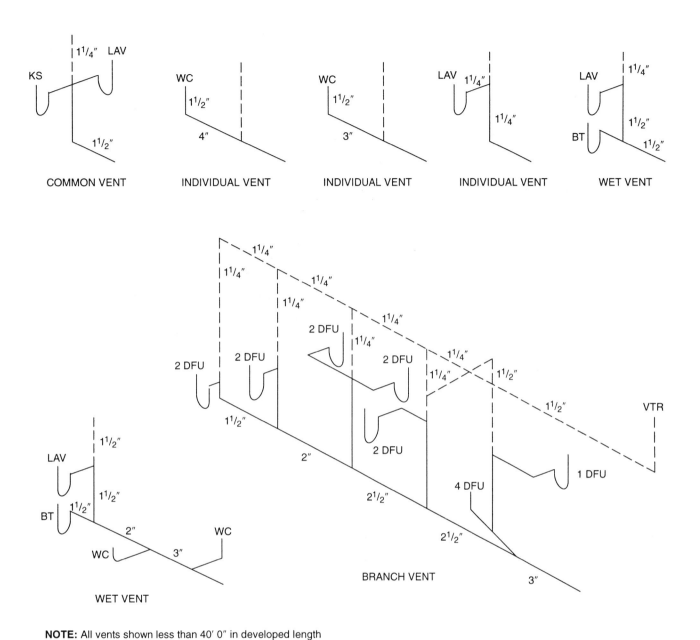

Figure 4-6 *Vents*

NOTE: All vents shown less than 40' 0" in developed length

49

Contractor's Guide to the Plumbing Code

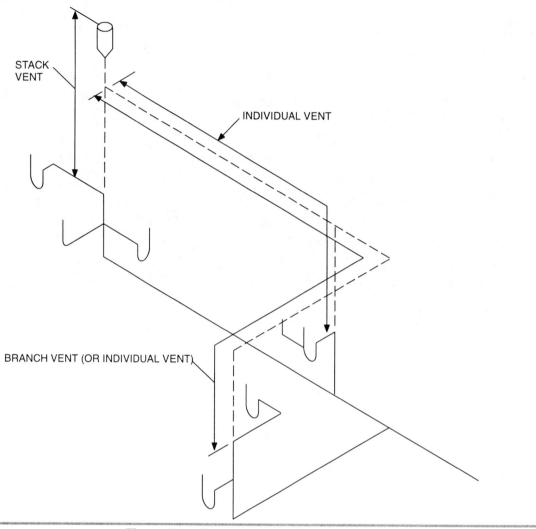

Figure 4-7 *Developed length of vents*

Conventional Drain, Waste and Vent Systems

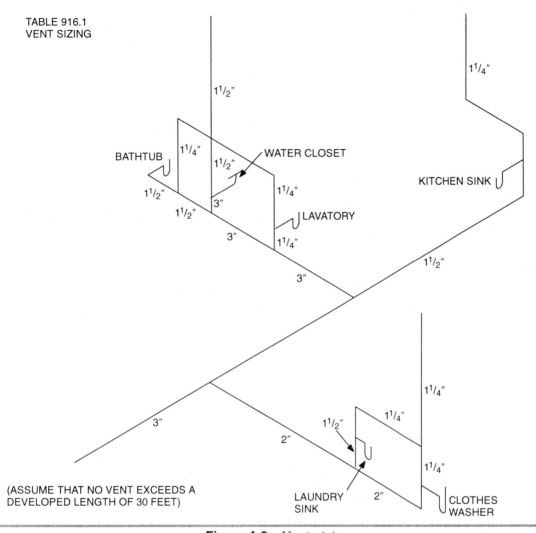

Figure 4-8 *Vent sizing*

Nonconventional Venting Methods

There are other approaches to venting that are outlined in the IPC. To some degree, based on your plumbing experience, the way these various venting methods work and how they are installed may appear difficult to understand. Keep in mind that if you are more familiar with a conventional venting system such as we just reviewed and as shown in Figure 4-8, then you may wish to stick with it until you become more familiar with the other methods. Nobody is twisting anyone's arm here to do things differently. A thorough and fair examination will show that a venting system designed in accordance with the other conventionally oriented codes such as the *1997 Uniform Plumbing Code*™, published by the International Association of Plumbing and Mechanical Officials, will, for the most part, meet the minimum requirements of the IPC. (See Appendix A concerning a comparison of the IPC and UPC.) In fact, with very few exceptions, this is true of the entire plumbing system design.

But don't be afraid to try something new. Who knows? You may find that these other venting methods can save you time and money and can be helpful in areas where the conventional venting method may be difficult to install.

Rather than review these methods in the order that they appear in the IPC, let's divide them into categories that can assist you in understanding the differences and similarities. There are a total of 10 methods of venting outlined in the IPC. Let's divide them into three categories:

1. Three specific venting methods, one of which is the conventional venting method already discussed;
2. Alternate venting methods; and
3. Venting methods where drains can serve as vents and vents can serve as drains.

We'll begin with the two remaining specific venting methods.

Island Fixture Venting (Section 913). This is a specific method for venting an island sink (see

Figure 4-9). It is limited to lavatories and sinks, except that a kitchen sink may have a food waste disposal and a dishwashing machine connected. The vent rises to a point above the drain outlet of the fixture before turning back down and connecting to a vertical drain pipe or to the top half of a horizontal drain. Prior to this connection, the vent shall branch off horizontally to a point where it can then rise vertically and either connect with another vent or extend to a proper termination point. Since the vent is below the flood rim of the fixture, it needs to be installed with drainage pattern fittings and be provided with a cleanout. The size of the vent (the dotted line in Figure 4-9) shall be one-half the diameter of the drain and, in no case less than a $1^1/_4$-inch diameter. As you look at the illustration you can see that this venting method provides for a free flow of air; liquid is not being trapped in the lowest portion of the vent because of its connection to the drain. Keep in mind though that this method requires considerable space to work with. You probably will not have enough room in a floor/ceiling joist space to install this vent properly. So you may wish to look at other venting methods. However, you'll find that it works well in a crawl space or underground application.

Relief Vents—Stacks for More Than 10 Branch Intervals (Section 914). This venting method requires a relief vent equal to the size of the vent stack it connects with for buildings exceeding 10 branch intervals, or at least 10 stories. In other words, you will include a vent stack that connects downstream of the waste stack in the top of the horizontal drain and rises vertically alongside the waste stack. The two stacks shall be connected as shown in Figure 4-10. This connection will occur at each tenth branch interval counting from the top. In addition, where the waste stack is offset and there are five or more branch intervals connected to the stack above the offset, the upper and lower sections of the offset shall be vented as shown in Figure 4-11.

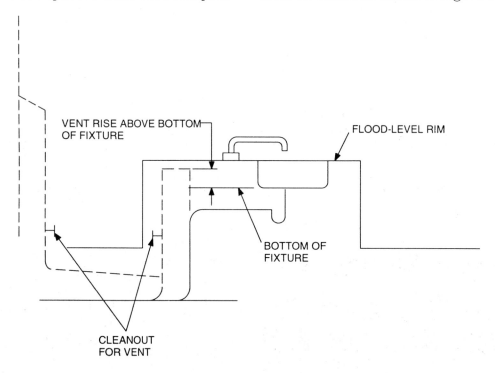

Figure 4-9 *Island fixture vent*

Conventional Drain, Waste and Vent Systems

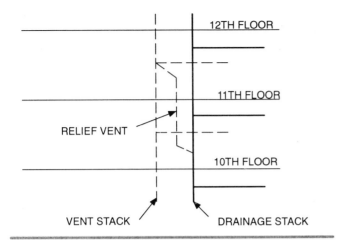

Figure 4-10 *Relief vent connection*

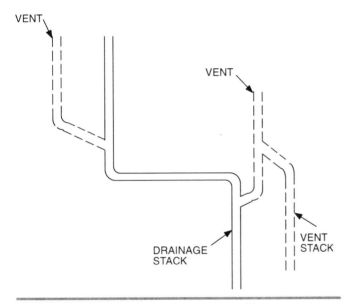

Figure 4-11 *Venting the offset*

Now let's briefly review two methods of venting that could be considered as alternates:

Air-admittance Valves (Section 917). The air-admittance valve (AAV) is a device designed to allow air into the vent pipe when negative pressures develop in the system. In this way, it is used on individual vents, branch vents and circuit vents in lieu of terminating to the exterior of the structure. Since the AAV will not allow positive pressures to pass through, there are certain installation requirements that are specified in the IPC. In some cases the IPC will be more restrictive than the AAV manufacturer's installation instructions. Thus you will want to give attention to the IPC requirements. AAVs are only allowed to vent fixtures that are on the same floor and that connect to a horizontal branch drain. They are not approved for waste stack vents, a venting method we will review later on in this chapter. The AAV essentially is the vent ter-

mination and therefore it must be installed within the maximum allowed developed length of the vent. The AAV must be a minimum of 4 inches above the horizontal branch drain or fixture drain being vented. Of course, access needs to be provided to the AAV, and since it draws air, it understandably has to be in a ventilated area. When located in an attic it needs to be at least 6 inches above the insulation. Also, make sure that the testing of the DWV system is completed prior to installing the AAV.

Probably the most important factor to remember is the relief vent. Recall that the AAV will only accommodate negative pressure. Any positive pressure that may result from the sewer, septic system or other portions of the DWV system must be allowed to exit the system by means of properly located relief vents. This is accomplished by having at least one vent extending to the free atmosphere. In cases where the horizontal branch connects to a stack with more than four branch intervals, a relief vent is required. The relief vent must connect to the horizontal branch drain between the stack or building drain and the most downstream fixture drain connected to the horizontal branch drain. Relief vents again must be at least one-half the diameter of the drain they serve, but in no case less than $1^1/_4$ inches in diameter (see Figure 4-12). The relief vent can also serve as a drain for a fixture.

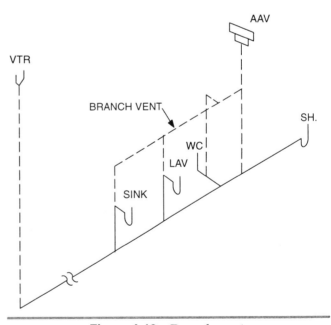

Figure 4-12 *Branch vent*

Now don't be alarmed! It may sound like you're installing a quadrillion vents when you are dealing with a building of more than four stories. But in reality, you can tie the relief vents together into one vent sized in accordance with Section 916.2. In this way you can end up with only one vent penetrating the exterior of the structure. Under certain types of construction this can save you some time, labor and materials. In larger commercial structures or multi-family buildings it may be prudent to seek the assistance of a professional designer.

Engineered Vent Systems (Section 918). This type of venting system is just that—an engineered system. Notice that according to Table 918.2, which provides criteria for determining cubic feet or air flow per minute for various diameters of pipe, a vent can be as small as $1/2$ inch in diameter. Well, guess what? You got it—no doubt you will need to consult a professional designer for this stuff and it will need to be submitted for review by the code official in accordance with Section 105.

Now that we have covered conventional venting, specific venting methods and alternate methods, let's look at the remaining venting methods for which the IPC is probably best known. As we review these methods, keep in mind that they are tested methods of venting that are used in various locations across the United States and elsewhere, and they work! These remaining five venting methods allow portions of drains to serve as vents and vice-versa. Let's examine them one by one.

Common Vent (Section 908). This very simple concept allows two traps to be vented by the same vent. The two traps and their fixture drains can be either at the same level or at different levels, provided that they are on the same floor. Figure 4-13 shows some of the various approaches to common venting. The sizing of common venting is specified in Table 908.3.

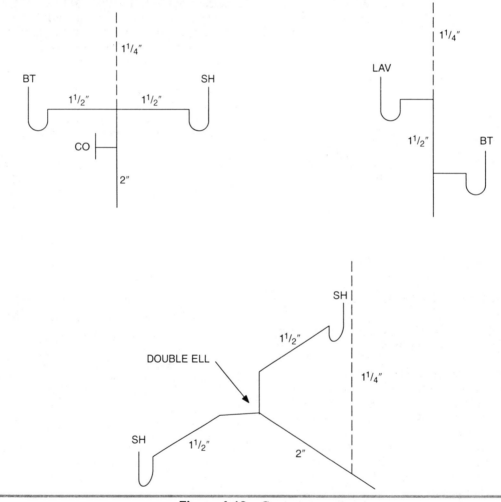

Figure 4-13 *Common vent*

Waste Stack Vent (Philadelphia Single-Stack) (Section 910). Again, this is a very simple method of allowing a drainage stack to serve as a vent. It greatly extends the concept of a vertical common vent discussed previously except for some specific guidelines for installation. For this system to function effectively there cannot be any offsets in the waste stack portion. Once the highest branch interval is connected to the stack, the vent portion may contain offsets, provided such offsets are at least

6 inches above the branch interval connection. The size of the waste stack vent is to be in accordance with Table 910.4. The thing to remember here is that the size of the stack, which is based on the total fixture units being drained into it by means of all branch intervals, shall be maintained the same size from the lowest point of the stack to the vent termination or connection to another approved vent. In other words, if the fixture unit total would require a 3-inch-diameter drain per Table 910.4, then both the stack and its vent must be 3 inches in diameter with no offsets at all until at least 6 inches above the highest branch interval (see Figure 4-14).

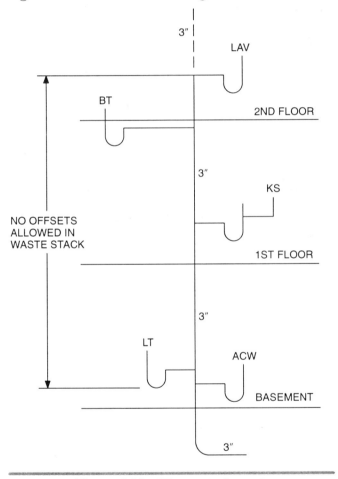

Figure 4-14 *Waste stack vent*

These two methods of using the drain as a vent, common venting and waste stack venting, are, for the most part, vertical in their application. Now let's examine the three remaining methods of such venting that are more often found in a horizontal application.

Wet Venting (Section 909). As you can see in Figure 4-15, the wet venting method can be used in both vertical and horizontal applications. The vertical method may be more common to plumbing installers in certain regions. However, don't be alarmed by what you see! The horizontal wet venting method is an effective way to provide a safe, sanitary system that might save you some time and materials.

Wet venting is limited to any combination of fixtures within two bathroom groups. Reviewing the definition for a bathroom group, we can ascertain that the wet vent system cannot exceed two water closets, two lavatories, two bidets, and two bathing facilities (bathtub, shower, or combination), for a total of eight fixtures, not including the two emergency floor drains that are also allowed. The two bathroom groups must also be located on the same floor level. The fixture drain length from the wet vent is limited to the distances shown in Table 906.1. The dry vent, the vent that continues upward from the wet vent to the point of termination or connection to another vent, shall be either an individual vent or a common vent to the lavatory, bidet, shower or bathtub, but not the water closet. In horizontal applications, the dry vent needs to be installed downstream of the first fixture drain or as an individual vent or common vent upstream of the last fixture drain. The dry vent must be sized in accordance with Section 916.2. Remember, though, that in this case, the drain served is going to be the entire wet-vented section. It will need to be at least one-half the diameter of the largest section of pipe in the system. The wet-vented section itself is sized per Table 909.3. Keep in mind that any additional fixtures outside of the wet vent that are to be drained into the same branch must be connected to the branch downstream of the wet vent. Since the wet vent serves as a drain, the type of fittings used must be in accordance with the drainage pattern shown in Table 706.3.

Circuit Venting (Section 911). This type of venting method is similar to wet venting except that it allows you to combine a total of eight fixtures on a single floor that are not limited to the two bathroom groups. It might be easier to explain by examining how circuit venting differs from wet venting. The fixture drains shall connect horizontally to the horizontal branch being circuit vented. Again the fixture drains are limited in length to those shown in Table 906.1. Since circuit venting is only to be used on horizontal applications, as opposed to wet venting, which can be both horizontal or vertical installations, the maximum slope for a circuit vent is 1 unit in 12 units horizontally, or 8-percent slope. The entire length of the circuit vent portion of the horizontal branch shall be sized for the total drainage discharge to the branch. There is not a unique sizing table for circuit venting as there is for wet venting or common venting (see Figure 4-16).

The circuit vent connection must be located between the two uppermost fixture drains and shall connect to the horizontal branch. It cannot serve as a drain for other fixtures—it is truly a dry vent. Where a circuit vent consists of four or more water closets, and discharges into a drainage stack that also receives the discharge of upper horizontal branches, a relief vent shall be connected to the horizontal branch ahead of the connection to the drainage stack and after the most downstream fixture drain of the circuit vent. Additional fixture drains may be connected with the circuit vent but they need to be vented by means other than the circuit vent, and the FUVs would be added to the total fixture-unit discharge into the horizontal branch to determine the size of the circuit vent. Such fixtures also must be located on the same floor as the circuit vent to which they connect (see Figure 4-17). Where the relief vent receives the discharge of other fixtures, the maximum discharge allowed is 4 drainage fixture units.

Combination Drain and Vent System (Section 912). This is a horizontal wet vent system limited to floor drains, standpipes, sinks and lavatories for the purpose of venting, except that a vertical riser, not to exceed 8 feet in length, may be used to connect a fixture drain to the horizontal combination drain and vent system. Again, the idea here is that the top half of the horizontal drain acts as a vent. As long as both the horizontal drain and vent system and the maximum 8-foot riser to a fixture drain are sized in accordance with Table 912.3, the flow of free air will be sufficient for the proposed design.

Of course, to get that free flow of air, a vent to the atmosphere must be provided. The vent, which must be sized for the total drainage fixture load of the combination drain and vent system per Section 916.2, can be located anywhere on the system and must rise vertically at least 6 inches before offsetting. This type of system is unique in that a branch that is already vented can accept a fixture drain under this combination drain and vent method (see Figure 4-18). Remember, too, that this type of venting procedure is dependent on adequate sizing and maintaining a horizontal installation. For this reason, the horizontal portion must not exceed a slope of $1/2$ unit vertical in 12 units horizontal, or 4-percent slope.

Now that wasn't too bad, was it? Many experienced plumbers will agree that the venting requirements of a plumbing system are usually the most difficult to comprehend. And since the IPC contains more types of venting methods than some other plumbing codes it can seem even more confusing. But as you continue to examine these methods and the illustrations over and over again, you will begin to comprehend them and will likely see how they can provide a viable alternative to what you probably are most familiar with. It's been said that if you hear something five times you are 20 percent more likely to recall it at a later date. This is very likely the case with venting techniques.

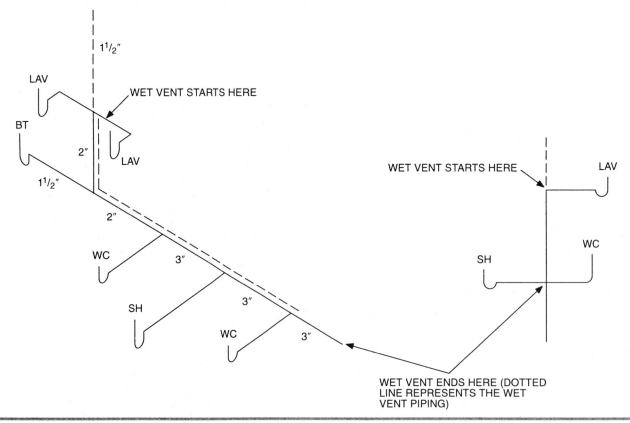

Figure 4-15 *Wet venting*

Conventional Drain, Waste and Vent Systems

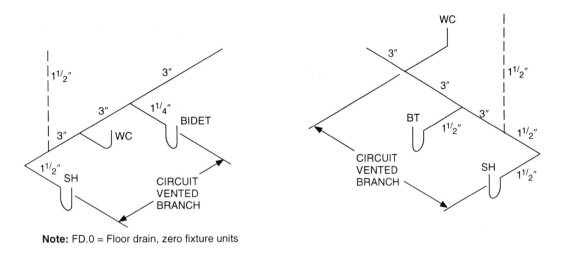

Note: FD.0 = Floor drain, zero fixture units

Figure 4-16 *Circuit venting*

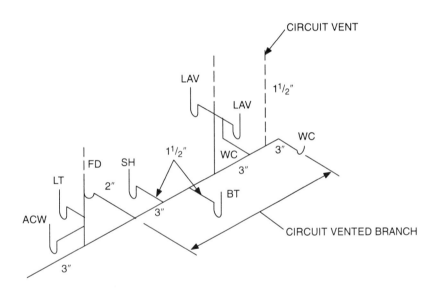

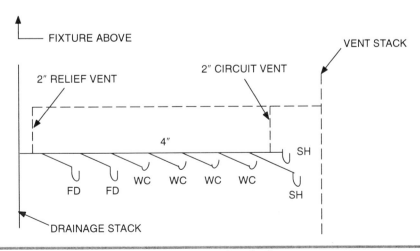

Figure 4-17 *Circuit vent with relief vent connection*

Contractor's Guide to the Plumbing Code

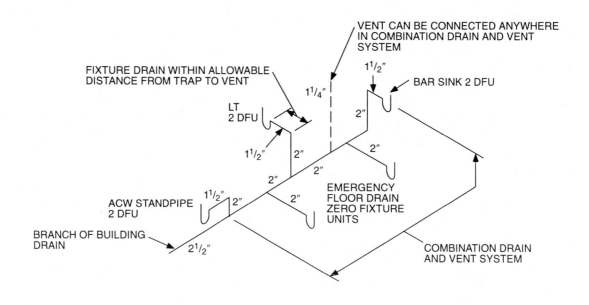

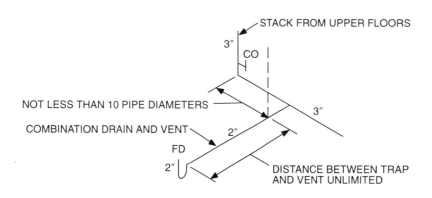

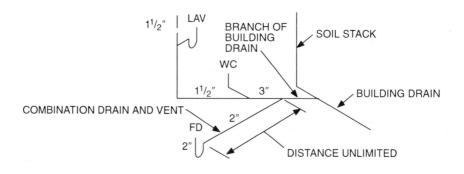

Figure 4-18 *Combination drain and vent*

Indirect Drains and Special Wastes

Most of the fixtures that you will be dealing with are what might be referred to as direct-connect fixtures. In other words, there is a tight seal between the fixture outlet and the drainage system. However, there are some types of fixtures that are considered indirect drain fixtures.

The most common indirect drain fixture in a residence would probably be the automatic clothes washer. As you know, a typical residential-type clothes washer has a drain hose that dips into a standpipe of the drainage system. Standpipes are required to be individually trapped and shall be between 18 inches and 42 inches in length above the trap. This provides enough height so as not to flood the drain to an overflow point due to the pumping action of the clothes washer.

More often than not you will be dealing with indirect drains in commercial applications, especially for health reasons, such as in food service establishments. Just as you want to prevent the odors and contents of the sewage disposal system from permeating your home, you would no doubt prefer to keep such odors and substances away from your New York Steak Special! Now you might immediately reason that an "indirect" drain, a drain that is not directly connected to the sanitary system, doesn't sound too sanitary at all. But, in reality, it is most sanitary when done correctly and is a must in protecting food during processing. To understand this concept, let's examine the essentials of an indirect drain.

Where a fixture is drained by means of an indirect drain, the receptor for the fixture must be an approved receptor—typically a floor sink, standpipe, hub drain, etc. A good rule of thumb is to consider what is being drained. If you are draining a sink, a floor sink is appropriate. If the drainage is being pumped, you might need to install a standpipe or a floor sink with a 3-inch trap rather than a 2-inch trap. A hub drain is effective for drains that release small quantities of liquid at a time. Remember that each receptor must be trapped to prevent sewer gases from escaping.

Here is where you'll need to determine whether the indirect drain will use an airbreak or an air gap (see Figure 2-2). Without a doubt, a physical gap between the drain pipe from a fixture/appliance and the flood rim of the receptor is the most sanitary. So if you're ever unsure about which one to use, you'll always be safe with the air gap. Otherwise, just remember that wherever the fixture or appliance can in any way siphon the contents of the receptor back to its potable water outlet, or where the fixture is to be used for the preparation of food, use an air gap. An air break is most effective for draining mechanical equipment or equipment that drains by means of pumping action and is provided with an air gap or other backflow device at its connection to the potable water system, such as a dishwasher or clothes washer (see Chapter 5).

Installing the Pipe and Fittings

Once you have determined the type of material you will use, you will need to follow the appropriate installation standards for joining the pipe and fittings and supporting the piping. Section 705 of the IPC outlines the various methods of joining pipe and fittings. In most cases it will not be difficult to determine the appropriate joining methods for the piping system you have chosen. Probably the most common materials used are acrylonitrile butadiene styrene (ABS) plastic, polyvinyl chloride (PVC) plastic, and cast iron. Simply put, use ABS fittings with ABS pipe, PVC fittings with PVC pipe, and cast-iron fittings with cast-iron pipe.

Of course it's not quite that simple. You need to join the materials together appropriately. Here again you will need to look for the properly listed materials to use. For instance, ABS pipe and fittings cannot be properly joined using PVC primers and cement, and vice-versa. Just because the advertisement says, "Glues anything, anytime, in any kind of weather," does not mean it is listed and approved for this specific use. So make sure that you follow the guidelines of Section 705 and use only the appropriate listed materials.

Along with the materials, you must familiarize yourself with the appropriate installation techniques. The latest IPC Commentary is a useful tool; it explains how to prepare the material for joining, from cutting, to reaming, to solvent welding. The integrity of the plumbing system depends on both the material and the joints. Trying to speed up the installation process by cutting corners will only serve to lessen the integrity of the system. Do it right the first time and you'll most likely avoid unnecessary repairs.

Table 308.5 provides the maximum spacing of pipe hangers when the piping is to be suspended horizontally or vertically. Of course, here again one needs to use pipe-hanging devices that are listed for the type of material. No, 14 gauge copper wire double-wrapped around a 16 d nail is not approved. Fast way to hang it? Yes! And it's a good thing that it's fast because you'll get to do it often—each time it fails!

Notice also how Table 308.5 works in conjunction with Section 308. Merely providing support at certain intervals may not adequately support the system where it is likely to experience vibration and movement during use. For this reason, criteria are provided to address the proper support at the base of stacks and at appropriate locations in horizontal runs to prevent swaying. Even more important is proper support and bracing for seismic purposes. This is an especially serious matter when installing materials such as cast iron.

Section 306 provides some guidelines for proper embedding and excavation for systems installed underground. Most important is the need to support the piping evenly so it will not move and settle. When placing the piping in an excavation, care must be exercised to ensure that there are no rocks or other materials that could damage the piping in any way. The problem here is that the damage may not reveal itself until years later. So be careful, and make sure the piping is being placed properly the first time.

Now that we have figured out how to drain the wastes out of the building, let's see how we can provide water to the building so our drains can go to work!

5

Water Supply and Distribution

The IPC regulates water piping systems in Chapter 6 and water heaters in Chapter 5. The requirements are consistent with accepted engineering practice for the design of water-distribution systems. One of the most important aspects of water-distribution systems is the protection against backflow. The IPC has extensive requirements that maintain the highest level of protection of the potable water system.

Water Is a Good Thing!

Potable water is water that is free from harmful impurities. It is the water we use to drink, process food, cook, and for bathing or showering. Water is so important to our health that we make rules and regulations about how it is used, what standards of quality it has to meet and how we will protect the future of this resource. We humans can survive much longer without food than we can without water. And since water is one of the best solvents known to man, the water-distribution system needs to be durable.

Ensuring a potable water supply is the fundamental undertaking of plumbing codes. This section will help guide you through the process of doing exactly that.

Water Services

Whether from a public service or private well, the water-service line to your structure has a minimum dimension of $3/4$-inch pipe, but further, it must be sized adequately to meet the total demand of the fixtures it serves.

When installing this service line, it must be a minimum of 12 inches below grade or 6 inches below the frost line in your area, as determined by the code official. For obvious reasons, if a sewer line is close by, there must be a minimum parallel separation of 5 feet, or the water line must be at least 12 inches above the sewer line. Potable water lines must not be installed in, under or above any potential sources of pollution; e.g., cesspools, septic tanks, leach lines or seepage pits. (See Chapter 8 for a discussion of these things.)

Sizing Those Pipes

Designing a water-distribution system can be done by using the tables listed in this code or by an engineer's design that meets with the approval of the code official. Table 604.3 lists flow rates and minimum flow pressures for the listed fixtures.

TABLE 604.3
WATER-DISTRIBUTION-SYSTEM DESIGN CRITERIA REQUIRED CAPACITIES AT FIXTURE SUPPLY PIPE OUTLETS

FIXTURE SUPPLY OUTLET SERVING	FLOW RATE[a] (gpm)	FLOW PRESSURE (psi)
Bathtub	4	8
Bidet	2	4
Combination fixture	4	8
Dishwasher, residential	2.75	8
Drinking fountain	0.75	8
Laundry tray	4	8
Lavatory	2	8
Shower	3	8
Shower, temperature controlled	3	20
Sillcock, hose bibb	5	8
Sink, residential	2.5	8
Sink, service	3	8
Urinal, valve	15	15
Water closet, blow out, flushometer valve	35	25
Water closet, flushometer tank	1.6	15
Water closet, siphonic, flushometer valve	25	15
Water closet, tank, close coupled	3	8
Water closet, tank, one piece	6	20

For SI: 1 psi = 6.895 kPa, 1 gallon per minute (gpm) = 3.785 L/m.

a. For additional requirements for flow rates and quantities, see Section 604.4.

The minimum pressure of the water service cannot be less than the minimum listed for the highest of any of the fixtures used. Example: If, in a single bathroom, a tank-type toilet, lavatory and shower are installed, the flow pressure listed for the shower, minimum 20 psi shown on Table 604.3, would be the minimum pressure allowed for that house because the shower required the most pressure. If the pressure provided by the water purveyor or private system is less than the minimum stated, a water pressure booster system would be required.

TABLE 604.4
MAXIMUM FLOW RATES AND CONSUMPTION FOR PLUMBING FIXTURES AND FIXTURE FITTINGS

PLUMBING FIXTURE OR FIXTURE FITTING	MAXIMUM FLOW RATE OR QUANTITY[b]
Water closet	1.6 gallons per flushing cycle
Urinal	1.0 gallon per flushing cycle
Shower head[a]	2.5 gpm at 80 psi
Lavatory, private	2.2 gpm at 60 psi
Lavatory (other than metering), public	0.5 gpm at 60 psi
Lavatory, public (metering)	0.25 gallon per metering cycle
Sink faucet	2.2 gpm at 60 psi

For SI: 1 gallon = 3.785 L, 1 gallon per minute = 3.785 L/m, 1 psi = 6.895 kPa.

a. A hand-held shower spray is a shower head.
b. Consumption tolerances shall be determined from referenced standards.

Table 604.4 lists the maximum flow rates and pressures for certain fixtures and fittings. In areas where the water pressure is higher than 80 psi, a pressure-reducing valve or regulator would be required. In commercial and institutional settings, there are many exceptions to this rule that require more water for proper use. (See Figure 5-1 and Figure 3-1.)

The illustration of a one-bedroom home is provided to give information regarding sizing of water piping using Tables 604.3, 604.4, 604.5 and 604.10.1. These generic tables are quick and easy to use.

We will start with a total fixture load to get the main water-line sizing, starting from the meter to the first branch. That would be a: clothes washer—4 gpm; laundry sink—4 gpm; lavatory—2 gpm; water closet (tank)—3 gpm; bathtub—4 gpm; kitchen sink—2.5 gpm; dishwasher—2.75 gpm; and hose bibb—5 gpm. (These figures are taken from Table 604.3.) Adding them together we get a total of 27.25 gpm.

The next step is to determine the pressure of the water supply. Table 604.10.1 gives a choice for maximum demand (gpm).

Proper sizing of the system is dependent on many variables, such as length and type of material, fittings and valves, elevation and regulators or filtering equipment, just to name a few. Appendix E outlines one such method for calculating the sizes required, which was used to determine the sizes shown in Figure 5-1.

It is important to note that this sizing method is based on the minimum pressure available, head changes in the system due to friction and elevation, and the rates of flow necessary for operation of various fixtures. This appendix will especially aid the larger fixture-unit installations. Note: Care must be taken when researching all pertinent data.

Table 604.5 lists the minimum pipe size for each fixture, with the exception of manifolds on parallel water systems that would allow reduction of that pipe by one size. Physical access to manifold systems is required, and each branch of the manifold is to be valved separately. Table 604.10.1 lists the manifold sizing, then displays the size of the pipe outlet on the manifold and gives the maximum water output for sizing the manifold itself. (See Figure 5-2 for an illustration of a manifold system.)

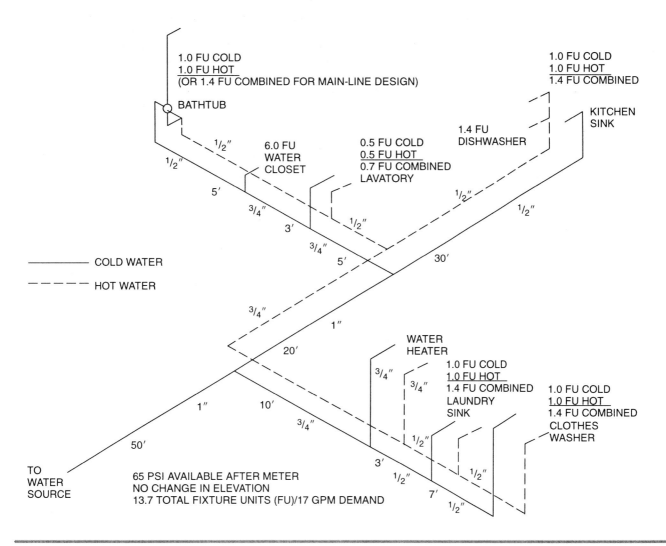

Figure 5-1 *Water distribution system*

TABLE 604.5
MINIMUM SIZES OF FIXTURE WATER-SUPPLY PIPES

FIXTURE	MINIMUM PIPE SIZE (inch)
Bathtubs (60" × 32" and smaller)[a]	1/2
Bathtubs (larger than 60" × 32")	1/2
Bidet	3/8
Combination sink and tray	1/2
Dishwasher, domestic[a]	1/2
Drinking fountain	3/8
Hose bibbs	1/2
Kitchen sink[a]	1/2
Laundry, 1, 2 or 3 compartments[a]	1/2
Lavatory	3/8
Shower, single head[a]	1/2
Sinks, flushing rim	3/4
Sinks, service	1/2
Urinal, flush tank	1/2
Urinal, flush valve	3/4
Wall hydrant	1/2
Water closet, flush tank	3/8
Water closet, flush valve	1
Water closet, flushometer tank	3/8
Water closet, one piece[a]	1/2

For SI: 1 inch = 25.4 mm, 1 foot = 304.8 mm, 1 psi = 6.895 kPa.

a. Where the developed length of the distribution line is 60 feet or less, and the available pressure at the meter is a minimum of 35 psi, the minimum size of an individual distribution line supplied from a manifold and installed as part of a parallel water-distribution system shall be one nominal tube size smaller than the sizes indicated.

**TABLE 604.10.1
MANIFOLD SIZING**

NOMINAL SIZE INTERNAL DIAMETER (inches)	MAXIMUM DEMAND (gpm)	
	Velocity at 4 feet per second	Velocity at 8 feet per second
1/2	2	5
3/4	6	11
1	10	20
1 1/4	15	31
1 1/2	22	44

For SI: 1 inch = 25.4 mm, 1 gpm = 3.785 L/m, 1 foot per second = 0.305 m/s.

Water Hammer

The latest edition of the IPC Commentary states that:

"Water hammer is a phenomenon that occurs from the dissipation of the kinetic energy of flowing water. The dissipation of energy occurs in the form of a pressure surge that produces a hydraulic shock wave traveling in excess of 3,000 miles per hour (1341 m/s). The pressure surge is sometimes accompanied by a banging noise from the piping system, hence the expression "water hammer." Any time flowing water is abruptly stopped, there is water hammer. For example, when a valve is closed, there is water hammer. To prevent any adverse effect from water hammer, the code requires the intensity of the water hammer to be controlled. Water hammer is more than simply an annoyance. The shock wave is accompanied by a pressure surge that expands the wall of the pipe. The pressure surge can result in damage to the piping system, fixture or water heater. The shock wave magnitude is generally smaller in plastic piping systems because the material absorbs some of the energy."

The intensity of water hammer is affected by the velocity of the flowing water, the rate at which the water flow is stopped, and the diameter and material of the pipe. As evidence, the critical factor is the velocity of flow. The intent of the code is to protect the entire water-distribution system and its components from the possible destructive forces that result from the rapid deceleration of water flow. For example, a shock wave having a magnitude of approximately 280 psi can occur in 1/2-inch-diameter copper tube when water having a velocity exceeding 5 feet per second is abruptly stopped, whereas the shock wave in 1/2-inch polybutylene pipe has a pressure of 34 psi at velocities exceeding 5 feet per second. The code requires the system to be designed for flow velocities that minimize the occurrence and magnitude of water hammer, and further requires water-hammer-arrester devices to be installed where quick-closing valves are to be used (see Commentary, Section 604.1).

Mechanical water-hammer arresters (shock arresters) help alleviate the water-hammer intensity by absorbing the energy (see Figure 5-3). When installed in areas where velocity cannot be adequately controlled, mechanical arresters should be located where they produce the maximum control of intensity. The manufacturer typically supplies guidelines for locating the arrester.

Because many water-hammer arresters are now permanently charged and factory sealed, they require no maintenance and should last the life of the system. Therefore, such devices may be installed in a concealed location without access when allowed per the manufacturer's instructions.

Note the distinction between water-hammer arresters and simple air chambers. An air chamber as shown in Figure 5-4 typically constructed of a capped section of vertical pipe, is ineffective in controlling water hammer. An air chamber located at the supply to every fixture was assumed to help control water hammer; however, air chambers became waterlogged in a short time. Air chambers allow a direct interface between the trapped air and the water, which allows the air to be absorbed or displaced in a short period of time (see the definition of "Quick-closing Valve").

The process of closing valves causes this problem and water-hammer arresters are required near quick-closing valves to solve it. Note: Properly strapped piping and padded supports also aid in stopping some additional vibration noises that can accompany water hammer.

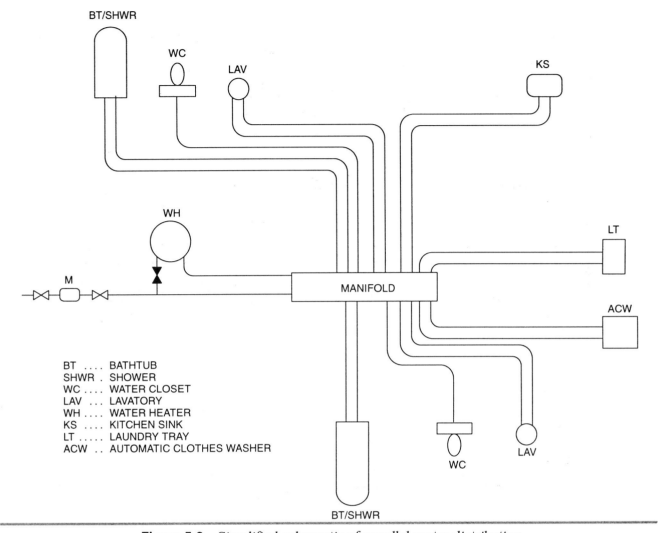

Figure 5-2 *Simplified schematic of parallel water distribution*

Contractor's Guide to the Plumbing Code

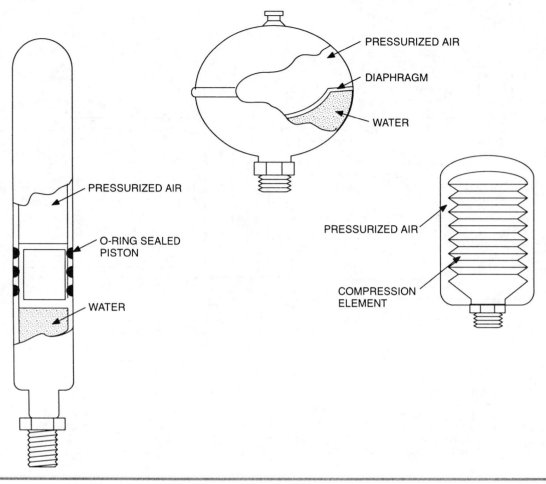

Figure 5-3 *Example of mechanical water-hammer arresters*

Water Supply and Distribution

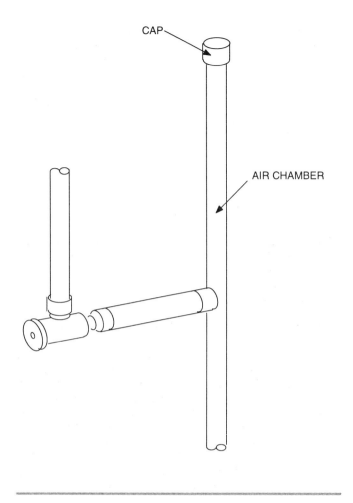

Figure 5-4 *Fixture supply*

TABLE 605.4
WATER-SERVICE PIPE

MATERIAL	STANDARD
Acrylonitrile butadiene styrene (ABS) plastic pipe	ASTM D 1527; ASTM D 2282
Asbestos-cement pipe	ASTM C 296
Brass pipe	ASTM B 43
Copper or copper-alloy pipe	ASTM B 42; ASTM B 302
Copper or copper-alloy tubing (Type K, WK, L, WL, M or WM)	ASTM B 75; ASTM B 88; ASTM B 251; ASTM B 447
Chlorinated polyvinyl chloride (CPVC) plastic pipe	ASTM D 2846; ASTM F 441; ASTM F 442; CSA B137.6
Ductile iron water pipe	AWWA C151; AWWA C115
Galvanized steel pipe	ASTM A 53
Polybutylene (PB) plastic pipe and tubing	ASTM D 2662; ASTM D 2666; ASTM D 3309; CSA B137.8
Polyethylene (PE) plastic pipe	ASTM D 2239; CSA CAN/CSA-B137.1
Polyethylene (PE) plastic tubing	ASTM D 2737; CSA B137.1
Cross-linked polyethylene (PEX) plastic tubing	ASTM F 876; ASTM F 877; CSA CAN/CSA-B137.5
Cross-linked polyethylene/aluminum/cross-linked polyethylene (PEX-AL-PEX) pipe	ASTM F 1281; CSA CAN/CSA B137.10
Polyethylene/aluminum/polyethylene (PE-AL-PE) pipe	ASTM F 1282; CSA CAN/CSA-B137.9
Polyvinyl chloride (PVC) plastic pipe	ASTM D 1785; ASTM D 2241; ASTM D 2672; CSA CAN/CSA-B137.3

Is Cheap Pipe the Best Way?

There may be times when installing a cheaper material will be just as good as using an expensive material or process. The environment in which the product is placed may influence the decision as to which type of material is best. The presence of any chemicals in the immediate area that could come in contact with the piping could cause its failure. Sunlight can damage some piping materials. This would result from excessive exposure to sunlight during construction. Some materials are prone to fire damage. The contents of the water supply should be analyzed to see if the piping could be weakened when certain additives are found in the water. Is the piping material exposed to physical damage? The pressure ratings of the piping material are very important. All piping must have less than 8 percent lead content. All these potentials must be addressed before piping is installed.

TABLE 605.5
WATER-DISTRIBUTION PIPE

MATERIAL	STANDARD
Brass pipe	ASTM B 43
Chlorinated polyvinyl chloride (CPVC) plastic pipe and tubing	ASTM D 2846; ASTM F 441; ASTM F 442; CSA B137.6
Copper or copper-alloy pipe	ASTM B 42; ASTM B 302
Copper or copper-alloy tubing (Type K, WK, L, WL, M or WM)	ASTM B 75; ASTM B 88; ASTM B 251; ASTM B 447
Cross-linked polyethylene (PEX) plastic tubing	ASTM F 877; CSA CAN/CSA-B137.5
Cross-linked polyethylene/aluminum/cross-linked polyethylene (PEX-AL-PEX) pipe	ASTM F 1281; CSA CAN/CSA-B137.10
Galvanized steel pipe	ASTM A 53
Polybutylene (PB) plastic pipe and tubing	ASTM D 3309; CSA CAN3-B137.8

**TABLE 605.6
PIPE FITTINGS**

MATERIAL	STANDARD
Acrylonitrile butadiene styrene (ABS) plastic	ASTM D 2468
Cast iron	ASME B16.4; ASME B16.12
Chlorinated polyvinyl chloride (CPVC) plastic	ASTM F 437; ASTM F 438; ASTM F 439
Copper or copper alloy	ASME B16.15; ASME B16.18; ASME B16.22; ASME B16.23; ASME B16.26; ASME B16.29; ASME B16.32
Gray iron and ductile iron	AWWA C110; AWWA C153
Malleable iron	ASME B16.3
Metal Insert Fittings Utilizing a Copper Crimp Ring SDR9 (PEX) Tubing	ASTM F 1807
Polyethylene (PE) plastic	ASTM D 2609
Polyvinyl chloride (PVC) plastic	ASTM D 2464; ASTM D 2466; ASTM D 2467; CSA CAN/CSA-B137.2
Steel	ASME B16.9; ASME B16.11; ASME B16.28

The fittings for these materials are of most importance when designing a water-distribution system because they too, present certain limitations. Soldered joints may be inappropriate when installing near combustible materials. Mechanical joints have to be tightened per the manufacturer's specifications. Solvent-cemented joints have to be pressurized with water or pressurized less than the desired working pressure for air products. Fused joints have to be uniformly melted and joined. Crimped joints require enough space in tight areas to allow the task of crimping. There is little tolerance for error when connecting threaded joints together. Flared joints have to be accessible. Some of these materials can be interconnected if approved fittings are used.

The Installation

Section 606 explains how this system is pieced together. It includes information on the required installation of full open valves on the water service at both sides of the meter, at the structure, on pressurized tanks or water heaters, and on branch lines in multiple units serving individual dwelling units.

Too Hot for Me . . . Hot Water

Amazing! Hot water for bathing, cooking, cleaning and clothes washing—all required to maintain our health. And, yes, it's a requirement that hot water be supplied to the left side of the fixture. Piping insulation is required on piping for recirculating systems. Thermal expansion devices are required on the downstream side of any pressure-reducing valve or backflow-prevention device. The installation of the water heater will be discussed in Chapter 6.

Backflow Devices

This is one of the most misunderstood sections in the code. First and foremost, the primary purpose of a backflow-prevention device is to prevent contaminants from going into the public or private potable water supply.

Systems requiring backflow-prevention devices include but are not limited to: filtration devices; closed systems with chemical additives; irrigation systems; enclosed medical or dental cleaning equipment; boilers and heat exchangers; beverage dispensers; portable cleaning equipment; fire sprinkler systems; and any nonpotable source water or agent. The key to understanding when backflow-prevention devices are required is to consider what the intended use of the equipment or process is, and determine if nonpotable water can safely flow back into the potable water supply.

Table 608.1 addresses the types of backflow-prevention devices and the application of those devices. Table 608.15.1 gives information on minimum distances required for air gaps. Table 608.17.1 gives required minimum distances from areas of possible contamination when drilling for private water systems.

TABLE 608.1
APPLICATION FOR BACKFLOW PREVENTERS

DEVICE	DEGREE OF HAZARD[a]	APPLICATION[b]	APPLICABLE STANDARDS
Air gap	High or low hazard	Backsiphonage or backpressure	ASME A112.1.2
Antisiphon-type water closet flush tank ball cock	Low hazard	Backsiphonage only	ASSE 1002 CSA CAN/ B125
Barometric loop	High or low hazard	Backsiphonage only	(See Section 608.13.4)
Reduced pressure principle backflow preventer	High or low hazard	Backpressure or backsiphonage Sizes $^3/_8''$ - 16"	ASSE 1013 AWWA C511 CSA CAN/CSA-B64.4
Reduced pressure detector assembly backflow preventer	High or low hazard	Backsiphonage or backpressure (Fire sprinkler systems)	ASSE 1047
Double check backflow prevention assembly	Low hazard	Backpressure or backsiphonage Sizes $^3/_8''$ - 16"	ASSE 1015 AWWA C510
Double check detector assembly backflow preventer	Low hazard	Backpressure or backsiphonage (Fire sprinkler systems) Sizes $1^1/_2''$ - 16"	ASSE 1048
Dual-check-valve-type backflow preventer	Low hazard	Backpressure or backsiphonage Sizes $^1/_4''$ - 1"	ASSE 1024
Backflow preventer with intermediate atmospheric vents	Low hazard	Backpressure or backsiphonage Sizes $^1/_4''$ - $^3/_4''$	ASSE 1012 CSA CAN/CSA-B64.3
Dual-check-valve-type backflow preventer for carbonated beverage dispensers/post mix type	Low hazard	Backpressure or backsiphonage Sizes $^1/_4''$ - $^3/_8''$	ASSE 1032
Pipe-applied atmospheric-type vacuum breaker	High or low hazard	Backsiphonage only Sizes $^1/_4''$ - 4"	ASSE 1001 CSA CAN/CSA-B64.1.1
Pressure vacuum breaker assembly	High or low hazard	Backsiphonage only Sizes $^1/_2''$ - 2"	ASSE 1020
Hose-connection vacuum breaker	High or low hazard	Low head backpressure or backsiphonage Sizes $^1/_2''$, $^3/_4''$, 1"	ASSE 1011 CSA CAN/CSA-B64.2
Vacuum breaker wall hydrants, frost-resistant, automatic draining type	High or low hazard	Low head backpressure or backsiphonage Sizes $^3/_4''$, 1"	ASSE 1019 CSA CAN/CSA-B64.2.2
Laboratory faucet backflow preventer	High or low hazard	Low head backpressure and backsiphonage	ASSE 1035 CSA B64.7
Hose connection backflow preventer	High or low hazard	Low head backpressure, rated working pressure backpressure or backsiphonage Sizes $^1/_2''$ - 1"	ASSE 1052
Spill-proof vacuum breaker	High or low hazard	Backsiphonage only Sizes $^1/_4''$ - 2"	ASSE 1056

For SI: 1 inch = 25.4 mm.

a. Low hazard—See Pollution (Section 202).
   High hazard—See Contamination (Section 202).

b. See Backpressure (Section 202).
   See Backpressure, Low Head (Section 202).
   See Backsiphonage (Section 202).

TABLE 608.15.1
MINIMUM REQUIRED AIR GAPS

FIXTURE	MINIMUM AIR GAP	
	Away from a wall[a] (inches)	Close to a wall (inches)
Lavatories and other fixtures with effective opening not greater than $1/2$ inch in diameter	1	$1^1/_2$
Sink, laundry trays, gooseneck back faucets and other fixtures with effective openings not greater than $3/4$ inch in diameter	$1^1/_2$	$2^1/_2$
Over-rim bath fillers and other fixtures with effective openings not greater than 1 inch in diameter	2	3
Drinking water fountains, single orifice not greater than $7/_{16}$ inch in diameter or multiple orifices with a total area of 0.150 square inch (area of circle $7/_{16}$ inch in diameter)	1	$1^1/_2$
Effective openings greater than 1 inch	Two times the diameter of the effective opening	Three times the diameter of the effective opening

For SI: 1 inch = 25.4 mm.

a. Applicable where walls or obstructions are spaced from the nearest inside edge of the spout opening a distance greater than three times the diameter of the effective opening for a single wall, or a distance greater than four times the diameter of the effective opening for two intersecting walls.

TABLE 608.17.1
DISTANCE FROM SOURCES OF CONTAMINATION TO PRIVATE WATER SUPPLIES AND PUMP SUCTION LINES

SOURCE OF CONTAMINATION	DISTANCE (feet)
Barnyard	100
Farm silo	25
Pasture	100
Pumphouse floor drain of cast iron draining to ground surface	2
Seepage pits	50
Septic tank	25
Sewer	10
Subsurface disposal fields	50
Subsurface pits	50

For SI: 1 foot = 304.8 mm.

Healthcare Plumbing

Because we call a hospital an emergency facility, the code requires a secondary water service in case of failure of one of the water services. Hot water must be supplied to each hospital fixture, and special fixtures, which require water heated to a higher temperature per the manufacturer's specifications, must receive the required hot-water supply. Vacuum breakers in a healthcare setting are required to have 6-inch-minimum air gaps above the flood-level rim instead of 1-inch air gaps for general plumbing installations.

Some other related items include backflow preventers—used with clinical hydrotherapeutic and radiological equipment that is supplied with water and discharges to the waste system. Sterilizers must be equipped with a feed line on the water-control valve when directly connected to the water system. Condensate traps must be provided with a water source for recharge and flushing.

6

Fixtures that Use Water

The IPC regulates water-piping systems in Chapter 6 and water heaters in Chapter 5. The requirements are consistent with accepted engineering principles and practices in water-distribution systems. Backflow, a very important issue, is the prevention of contaminated water from entering the water-distribution system for which the IPC maintains the highest level of protection. Materials, sizing criteria, flow rates, fixtures and backflow are the important issues to be considered. These were discussed in the previous chapter. Now we will give attention to the final link in the distribution system—the fixture.

Fixtures

All the effort to make a quality installation of the drainage, venting and water-distribution system will be of little value if the fixtures do not comply with an approved, nationally accepted standard. This is where Chapter 4 of the IPC can help. It provides information as to the types and quality of fixtures that may be used, along with installation requirements to ensure that the fixtures will function properly and will be accessible for use.

As with other materials, fixtures must be listed and approved for use. Chapter 14 of the IPC will assist you in determining what fixtures can be used. Our discussion in Chapter 3 regarding the plastic shower enclosure test will help in understanding the testing criteria for fixtures. The main thing is that fixtures need to be self-scouring. This requires surfaces that will easily drain clean. Appropriate testing standards will address this aspect. Of course, this doesn't mean that you never need to do some cleaning yourself. Even the best quality fixture will prove "undesirable" in time if it is not maintained for cleanliness.

Number of Fixtures

IPC Table 403.1 provides some guidelines for the proper number and distribution of fixtures for various occupancy classifications. You will notice that this table requires considerably more water closets for female use than for male use, especially for assembly uses. There's a good reason for this. Somewhere on the road to modern plumbing design, no doubt, engineers finally listened to their spouses who were severely displeased with the need to stand in line for eternity to use the rest room at half-time of a ball game. As they viewed the second half of the game from a small television screen 20 feet in the air, their husbands quickly strolled through the rest room, took care of business, grabbed another drink, caught up on the re-cap of the first half on their headsets and made it back to their seats before half-time ended. The conversation on the way home from the game must have really sunk in!

Of course, there are occupancies where separate facilities would not be necessary, such as residences, motels and hotels, or occupancies with 15 or fewer occupants. Sections 403.4 and 403.5 provide criteria for determining where employee facilities are to be provided, while Section 403.6 addresses customer facility location.

Section 404 is important because it will give us the needed dimensions to ensure that an adequate number of fixtures are accessible to persons with disabilities. Most often, we tend to think of "wheelchair" accessibility when we consider this subject. However, accessibility can encompass a much wider range of disabilities beyond what may constitute the need for a wheelchair. It may involve a person's ability to see, feel, reach, push, pull, or various other motor skills. In fact, all of us can be thought of as "temporarily able-bodied." How many of us have broken a leg or arm playing, let's say, a sport that is better left to younger people? This section requires compliance with accessibility requirements of the *International Building Code*. If your construction project is such that these requirements are necessary, you will want to become familiar with accessibility aspects of the building code. Figures 6-1 and 6-2 show some of the basic access requirements. Figure 6-3 shows minimum distance requirements for standard installations not required to be accessible.

As with the rest of your plumbing installation, care must be exercised to ensure that the system can be maintained for sanitary purposes. With regard to fixtures, this would involve properly setting, securing and sealing the fixture. In no way do you want to set the fixture too close to surrounding walls so that they create surfaces that cannot be cleaned. It should go without saying that fixtures are to be set level and in proper alignment to adjacent walls, but we'll emphasize it anyway. And another thing (if we sound like we're harping at you, please excuse us), seal the joints where the fixture comes in contact with floors, walls or counters. This particular requirement of Section 405.5 is very important in preventing the growth of bacteria, not to mention protecting the integrity of the structure. It's wonderful to know that you can clean around the fixture. But even clean water that may enter areas between the fixture and its mounting surface can breed bacteria that cause structural deterioration. We have heard many a plumber say that you shouldn't seal a water-closet base to the floor. They claim that if the fixture leaks you want to allow a place for the leakage to exit from beneath the fixture. The bottom line is you're supposed to install it so it doesn't leak! And it makes sense to test a fixture to make sure there are no leaks before sealing the joints.

Another area of concern is the matter of making sure that the fixture stays put! You must adhere to the manufacturer's instructions for securing the fixture. Where a fixture hangs on a wall, you need to use the proper carrier that, again, must be installed per manufacturer's instructions. Remember, gypsum board does not make a good support for a wall-hung sink or urinal. Carriers are designed in such a way that the fixture can be supported. It also allows properly fitted drain connections that will not result in the connection carrying the weight of the fixture. The carrier is a manufactured device fastened to the structure to support the fixture, and in most cases will accompany the fixture at purchase. These must be installed as required by the manufacturer's specifications. Floor-mounted water closets also must be properly supported on a level floor space and attached securely to a flange. The flange in turn must be properly secured to the floor. Sections 405.4 through 405.4.3 give direction on these matters.

Where slip-joint connections are used to connect a fixture to the plumbing system, access to the connections must be provided. For lavatories this will be accomplished by means of cabinet doors beneath the sink. However, where the fixture connection is not located in a cabinet, such as a bathtub, you need to provide a minimum 12-inch by 12-inch access panel. For a bathtub, the access panel could appropriately be located in a ceiling directly below the bathtub. Thus it makes sense that access from an accessible underfloor space (crawlspace) would not need to have an additional access panel in the wall above the floor.

Fixtures that Use Water

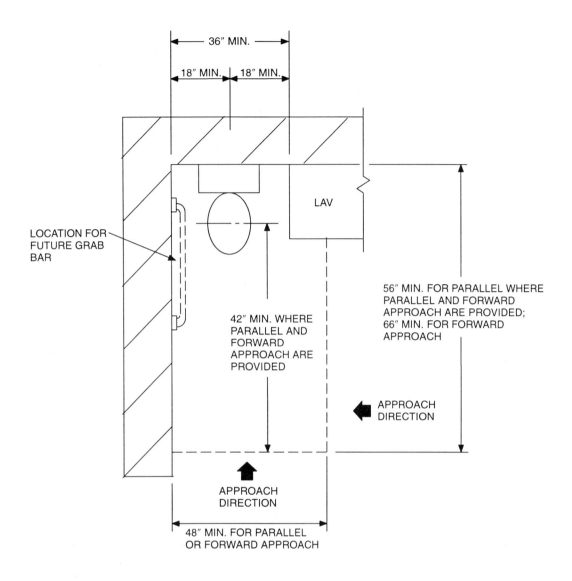

Figure 6-1 *Clear floor space for water closet in a Type B dwelling unit*

While, obviously, we are sure that none of our readers are what the trade affectionately calls "chainsaw plumbers," we do need to point out a couple of potential pitfalls.

Removing too much of the framing members or diaphragm to install plumbing pipes and fixtures can obviously result in structural failure. Also, such a cutout can form a passage through which a hidden fire could spread from one level to the next. *International Building Code* Chapter 7 requires fire blocking in combustible (wood) construction. Which trade's responsibility it is to do this is undefined, but it's more than likely that the building inspector will point out this critical feature.

Also in protected construction—types of construction II-B, III-B, IV and V-B—the floor membrane is the critical last barrier against premature failure of construction that should resist a fire's spread for a minimum of 60 minutes. In this case a much stricter standard than fire blocking is required for resistance to the spread. In fact, Section 106.3.1 requires that the designer state on the plans how such potential breaches are to be repaired. Suffice it to say that less than quality workmanship by the plumbing contractor can have life-and-death consequences for the occupants of the building. (Also see IPC Sections 305.5 and 307.3 for through-penetration fire-stop requirements, along with Chapter 7 of the IBC.)

Other Fixtures

The remaining portions of Chapter 4 contain lists and details of appropriate standards for various fixtures and appliances (see Sections 406 through 419). Just a couple of fixtures need to be addressed.

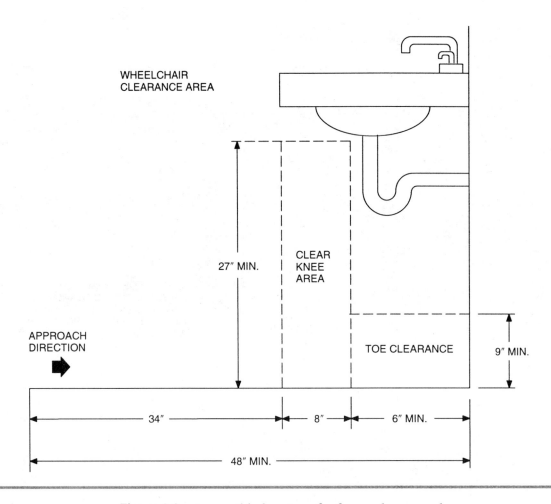

Figure 6-2 *Accessible lavatory for forward approach*

Dishwashers and Showers

Dishwashers are directly connected to the water supply and must be installed in a manner that prevents backflow of the internal contents into the water system. This can be accomplished by means of installing a backflow preventer or an air gap. However, if the dishwasher conforms to ASSE 1006, such backflow protection will already be provided. In all cases the drain for the dishwasher will need to connect to a drainage air gap (see Figure 6-4).

Showers are a great asset in maintaining one's personal hygiene. However, a shower stall is of little value if you can't get in it! Thus, the IPC provides minimum dimensions along with other size requirements in Section 417 (see Figure 6-5). Now let's see what is needed in order to avoid a cold shower.

Water Heaters [C-B]

Most of what we have discussed so far provides for comfortable and sanitary living. Since everyone appreciates the comfort of a warm shower or bath, installing a water heater that functions properly, especially on a cold day, is essential! However, in providing the necessary hot water, we must now examine the water heater, which can be one of the most, if not the most, unsafe devices when improperly installed. If such an improperly installed water heater explodes, it can become a miniature space shuttle that will launch from its pad in the basement and exit the roof on its maiden voyage toward the stars. But, as much as we may love space exploration, let's keep such launches where they belong and not in our house!

One of the first things we will want to do is to ensure that the water heater is listed and approved. The label will tell us the necessary facts about the standard it meets (the details of which are found in IPC Chapter 14), along with pressure ratings, proper clearances from combustibles and other pertinent information. Along with the information on the label, you will want to become familiar with the installation instructions. Remember though that the minimum provisions of the IPC must be met. If the installation instructions are less restrictive than the IPC, the IPC provisions will take precedence.

Fixtures that Use Water

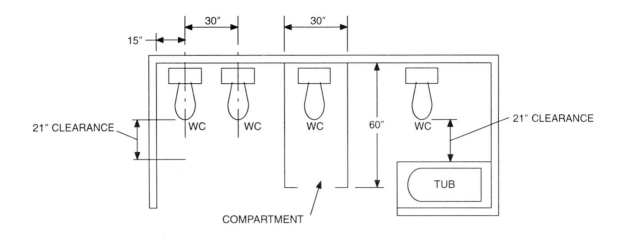

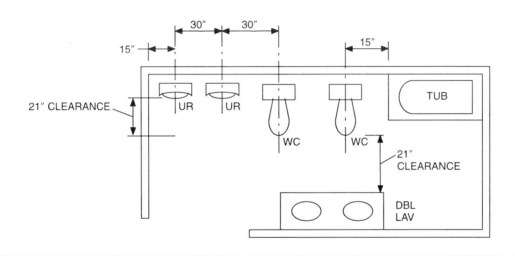

Figure 6-3 *Fixture clearances*

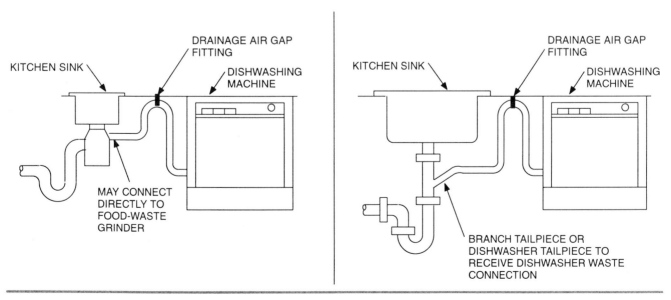

Figure 6-4 *Dishwasher waste connections*

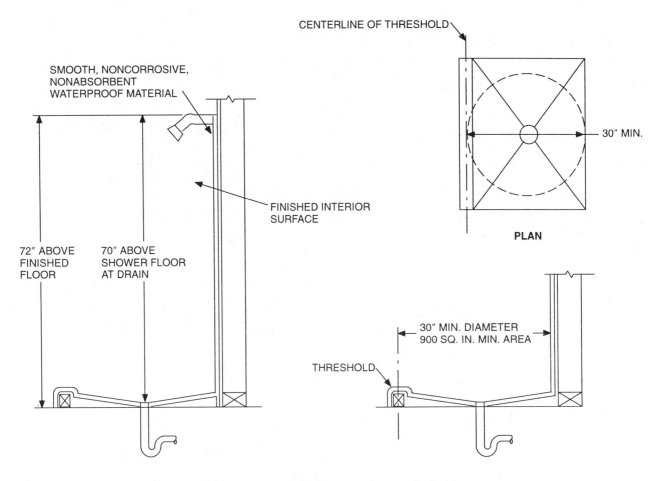

NOTE: Shower valves, grab bars, soap dishes, etc., are permitted to encroach on required minimum area.

Figure 6-5 *Shower enclosure*

Care must be exercised in locating the water heater. One important concern is when the water heater is located in a garage. In this situation the water heater must be elevated so that any ignition source will be at least 18 inches above the garage floor level. This applies to both fuel-fired and electric water heaters. Where the water heater is elevated you will need to give special attention to its support and bracing. There are a number of methods and/or products that can be used for this purpose. Be aware that if you live in an earthquake seismic zone there will be some specific provisions for bracing the installation that will apply. Talk with your code official about this and don't forget to check the manufacturer's specifications. And while we're on the subject of safety, since you're going to go the distance to secure the water heater in place, consult the *International Mechanical Code*® or *International Fuel Gas Code*® for adequate protection from vehicle impact (Figure 6-6).

Another location that will require some careful thought is when the water heater is installed in an attic. While it certainly seems like a nice place to tuck a water heater out of the way, you don't want yourself or the service technician to fall through the ceiling! That's why Section 502.5 provides some specific access requirements. The distance from the access opening to the water heater shall not exceed 20 feet. You must provide a continuous solid floor of at least 24 inches in width from the opening to the water heater, with a 30-inch by 30-inch working platform at the control area. Make sure also that the access opening is a minimum 20 inches by 30 inches, or is large enough to allow removal of the water heater.

Remember that while you might be able to set the water heater into place by yourself, you most likely will not be able to perform the same feat when the heater is full of water. If you have a 40-gallon water heater, you will be looking at about 300 pounds of water, along with the weight of the water heater. Check with your code official, truss or ceiling joist designer, building contractor, and any other appropriate individuals to determine if the water heater will be provided with adequate support.

That brings us to another issue. While you wish the water to remain in the water heater until you draw hot water for whatever reason, you must be prepared in the event that the water heater develops a leak. Since water and building contents don't mix very well, you will want to provide a drain pan beneath the water heater. Actually, this is a requirement found in Section 504.7. Let's discuss the size and installation of the drain pan.

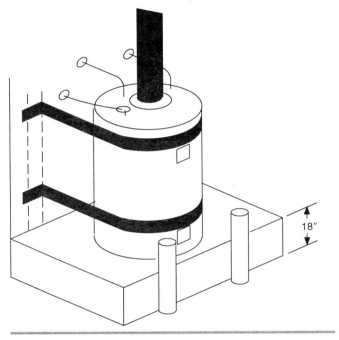

Figure 6-6 *Water Heater Strapping*

Drain pans can be either plastic or metal. However, if the pan is serving a gas water heater you will need to use a metal drain pan. The pan is required to be at least $1^1/_2$ inches deep and shall be equipped with a minimum 1-inch drain. The gravity drain must terminate outside the structure between 6 inches and 24 inches above ground, or it may terminate at a floor drain, receptor or other location approved by the code official.

Probably the most important part of the water heater installation is the temperature and/or relief valve conforming to ANSI Z21.22. The water heater must be equipped with a means of relieving pressure, thus both high temperature and excessive pressure must be addressed. In most cases this will be accomplished by means of a combination temperature- and pressure-relief valve. The settings shall not exceed 210°F or 150 psi of pressure. The sensing element of the relief valve must be within the upper 6 inches of water in the tank. In most cases the water heater will have an opening provided on its top or upper side.

The discharge for the relief line (see Section 504.6) must be no smaller than the size of the outlet on the relief valve. It must be installed so as to drain by gravity to the exterior of the building, or to an approved drain, receptor or other location as allowed by the code official. There is a distinct possibility that the outlet will discharge water at an extremely high temperature and pressure, so care must be exercised in determining how and where to terminate the drain in order to avoid personal injury or property damage.

You will need to install an air gap on the drain prior to its leaving the room or enclosure in which the water heater is located (see Figure 6-7). You need to prevent the drain from freezing and plugging up. Further, you must further install the drain in such a manner that the drain termination can be readily visible so as to alert occupants that servicing is needed. Thus you will want to avoid isolated points of termination, or terminations to a floor drain or floor sink that could appear to be a "normal functioning" drain from some other fixture, appliance or device.

Now that we've covered all the requirements for getting water into and out of the building, let's focus on some aspects of the IPC that will prevent rain and ground water from getting into the structure.

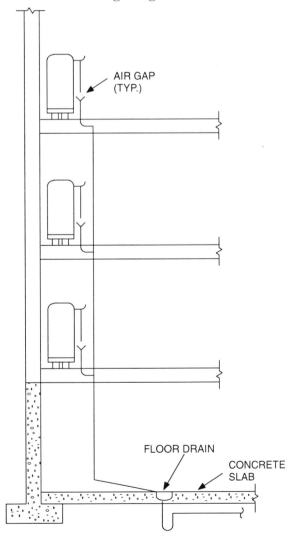

Figure 6-7 *Relief valve indirect waste*

7

Storm Drainage

General

Roof drains are required to direct rainwater off the roof and away from the structure. The roof drains we are referring to are more than gutters and downspouts. Although the IPC provides criteria for the installation of gutters and downspouts, we will discuss drains that are necessary for directing rainwater off buildings with flat roofs. In these cases the roof-drain system will not really be much different from the rest of the building's plumbing.

Storm drainage is also required for roofs, paved areas, yards, courts and courtyards. Drainage in general needs to be directed to approved locations, temporary storage devices, or public and private drainage systems. One- and two-family dwellings can discharge directly onto flat areas or streets, when approved, provided that the storm water flows away from the building. (See Section 3303.5 in the *International Building Code*.)

Sizing the system is really simple. Table 1106.2 provides the total roof area in square feet that can be drained with specific diameters of drain leaders. This is based on the rainfall rate in inches per hour. Figure 1106.1 will assist you in determining the rainfall rate for your location. To determine the horizontal piping sizes, use Table 1106.3 in the same way that you used Table 1106.2 for vertical leaders. The reader will note here that by using different grades of slope for the horizontal piping you are provided with numerous options for designing an effective and safe roof-drain system.

Materials

Determining the type of material to use is fairly easy. Tables 1102.4, 1102.5 and 1102.7 are straightforward and allow for quite a variety of materials to choose from. The reader should review definitions to be able to distinguish the difference between a storm drain, a subsoil drain, a leader, a conductor, and so on.

**TABLE 702.1
ABOVE-GROUND DRAINAGE AND VENT PIPE**

MATERIAL	STANDARD
Acrylonitrile butadiene styrene (ABS) plastic pipe	ASTM D 2661; ASTM F 628; CSA B181.1
Brass pipe	ASTM B 43
Cast-iron pipe	ASTM A 74; CISPI 301; ASTM A 888
Coextruded composite ABS DWV sch 40 IPS pipe (solid)	ASTM F 1488
Coextruded composite ABS DWV sch 40 IPS pipe (cellular core)	ASTM F 1488
Coextruded composite PVC DWV sch 40 IPS pipe (solid)	ASTM F 1488
Coextruded composite PVC DWV sch 40 IPS pipe (cellular core)	ASTM F 1488
Coextruded composite PVC IPS - DR, PS140, PS200 DWV	ASTM F 1488
Copper or copper-alloy pipe	ASTM B 42; ASTM B 302
Copper or copper-alloy tubing (Type K, L, M or DWV)	ASTM B 75; ASTM B 88; ASTM B 251; ASTM B 306
Galvanized steel pipe	ASTM A 53
Glass pipe	ASTM C 1053
Polyolefin pipe	CSA CAN/CSA-B181.3
Polyvinyl chloride (PVC) plastic pipe (Type DWV)	ASTM D 2665; ASTM D 2949; ASTM F 891; CSA CAN/CSA-B181.2; ASTM F 1488
Stainless steel drainage systems, types 304 and 316L	ASME/ANSI A112.3.1

**TABLE 706.3
FITTINGS FOR CHANGE IN DIRECTION**

TYPE OF FITTING PATTERN	CHANGE IN DIRECTION		
	Horizontal to vertical	Vertical to horizontal	Horizontal to horizontal
Sixteenth bend	X	X	X
Eighth bend	X	X	X
Sixth bend	X	X	X
Quarter bend	X	X[a]	X[a]
Short sweep	X	X[a,b]	X[a]
Long sweep	X	X	X
Sanitary tee	X[c]		
Wye	X	X	X
Combination wye and eighth bend	X	X	X

For SI: 1 inch = 25.4 mm.

a. The fittings shall only be permitted for a 2-inch or smaller fixture drain.
b. Three inches and larger.
c. For a limitation on double sanitary tees, see Section 706.3.

The design and materials used for storm drainage systems are similar to those used for drainage/waste systems because their principle use is much the same. Only the contents and disposal methods differ. Storm drainage piping for roof- and deck-drain systems located within the building must conform to the materials listed in Table 702.1.

The fittings applied to these materials must use the approved drainage-type fittings listed in Table 706.3. Cleanouts that are required will be installed using the same standards as those used in the sanitary system.

Roof Drain and Overflow Design

The roof drain itself will be level with the surface of the roof and shall be equipped with a strainer. Roof drains shall have strainers extending not less than 4 inches above the surface next to the roof drain. The allowable inlet area of the strainer above the roof can't be less than one and one-half times the area of the conductor or leader to which the drain is connected. In other words, if you have a 3-inch drain (7.06 square inches of area), the area of the strainer must be a minimum of 10.59 square inches. This requirement in combination with the 4-inch height requirement for the strainer will allow drainage when the strainer becomes partially clogged with debris.

Roof-drain strainers on sun decks or parking decks shall be of a flat surface design and must have not less than two times the area surface as compared to the conductor or leader to which the drain is connected. Roof-drain flashing needs to have approved flashing material and must be watertight.

Another protection is the requirement for secondary drainage. This is a must where water can accumulate on the roof due to parapet walls or similar construction should the roof drain become fully obstructed. This can be accomplished in a couple of ways. One is by providing a secondary drain independent of the primary drain. This will result in an increase in the size of the secondary drain in comparison with the primary. Another way to accomplish this is to provide scuppers through the parapet.

Secondary drains or scuppers should drain into a location where the building occupants will observe spilling water. This way, if they notice rain water draining from the secondary system, they will be alerted to the need for servicing the primary roof drain that has become inoperable.

Secondary drains shall be sized by the same methods as the primary roof drains on Tables 1106.2, 1106.3 and 1106.6. Overflow scuppers must be sized to prevent the depth of ponding water from exceeding that for which the roof was designed and shall have a minimum opening dimension of 4 inches (see Figure 2-10).

TABLE 1106.2
SIZE OF VERTICAL CONDUCTORS AND LEADERS

DIAMETER OF LEADER (inches)[a]	HORIZONTALLY PROJECTED ROOF AREA (square feet)											
	Rainfall rate (inches per hour)											
	1	2	3	4	5	6	7	8	9	10	11	12
2	2,880	1,440	960	720	575	480	410	360	320	290	260	240
3	8,800	4,400	2,930	2,200	1,760	1,470	1,260	1,100	980	880	800	730
4	18,400	9,200	6,130	4,600	3,680	3,070	2,630	2,300	2,045	1,840	1,675	1,530
5	34,600	17,300	11,530	8,650	6,920	5,765	4,945	4,325	3,845	3,460	3,145	2,880
6	54,000	27,000	17,995	13,500	10,800	9,000	7,715	6,750	6,000	5,400	4,910	4,500
8	116,000	58,000	38,660	29,000	23,200	19,315	16,570	14,500	12,890	11,600	10,545	9,660

For SI: 1 inch = 25.4 mm, 1 square foot = 0.0929 m².

a. Sizes indicated are the diameters of circular piping. This table is applicable to piping of other shapes, provided the cross-sectional shape fully encloses a circle of the diameter indicated in this table.

TABLE 1106.3
SIZE OF HORIZONTAL STORM DRAINAGE PIPING

SIZE OF HORIZONTAL PIPING (inches)	HORIZONTALLY PROJECTED ROOF AREA (square feet)					
	Rainfall rate (inches per hour)					
	1	2	3	4	5	6
$1/8$ unit vertical in 12 units horizontal (1-percent slope)						
3	3,288	1,644	1,096	822	657	548
4	7,520	3,760	2,506	1,800	1,504	1,253
5	13,360	6,680	4,453	3,340	2,672	2,227
6	21,400	10,700	7,133	5,350	4,280	3,566
8	46,000	23,000	15,330	11,500	9,200	7,600
10	82,800	41,400	27,600	20,700	16,580	13,800
12	133,200	66,600	44,400	33,300	26,650	22,200
15	218,000	109,000	72,800	59,500	47,600	39,650
$1/4$ unit vertical in 12 units horizontal (2-percent slope)						
3	4,640	2,320	1,546	1,160	928	773
4	10,600	5,300	3,533	2,650	2,120	1,766
5	18,880	9,440	6,293	4,720	3,776	3,146
6	30,200	15,100	10,066	7,550	6,040	5,033
8	65,200	32,600	21,733	16,300	13,040	10,866
10	116,800	58,400	38,950	29,200	23,350	19,450
12	188,000	94,000	62,600	47,000	37,600	31,350
15	336,000	168,000	112,000	84,000	67,250	56,000
$1/2$ unit vertical in 12 units horizontal (4-percent slope)						
3	6,576	3,288	2,295	1,644	1,310	1,096
4	15,040	7,520	5,010	3,760	3,010	2,500
5	26,720	13,360	8,900	6,680	5,320	4,450
6	42,800	21,400	13,700	10,700	8,580	7,140
8	92,000	46,000	30,650	23,000	18,400	15,320
10	171,600	85,800	55,200	41,400	33,150	27,600
12	266,400	133,200	88,800	66,600	53,200	44,400
15	476,000	238,000	158,800	119,000	95,300	79,250

For SI: 1 inch = 25.4 mm, 1 square foot = 0.0929 m².

Specific Design Criteria and Details

The parameters used to calculate leader or conductor sizes and for number of drains used are found in Tables 1106.2 and 1106.6.

But the first step in finding the minimum number is to look at the maximum rainfall average for your locality on the map of the United States found in Figure 1106.1 of the 2000 IPC (or approved data in your area). This number gives you the statistical projected amount of water accumulating in 1 hour during the 100-year rainfall event. We recommend this number be rounded up to the next whole number.

Using Tables 1106.2 and 1106.3, the rainfall rate for the area is determined. The flat surfaces that collect storm water are calculated at 100 percent. One-half the area of vertical walls that divert water onto the roof must be added in the square footage of that area served (see Section 1106.4). Add both of these numbers—one-half the total vertical wall area and the horizontal square footage.

TABLE 1106.6
SIZE OF SEMICIRCULAR ROOF GUTTERS

DIAMETER OF GUTTERS (inches)	HORIZONTALLY PROJECTED ROOF AREA (square feet)					
	Rainfall Rate (inches per hour)					
	1	2	3	4	5	6
$1/16$ unit vertical in 12 units horizontal (0.5-percent slope)						
3	680	340	226	170	136	113
4	1,440	720	480	360	288	240
5	2,500	1,250	834	625	500	416
6	3,840	1,920	1,280	960	768	640
7	5,520	2,760	1,840	1,380	1,100	918
8	7,960	3,980	2,655	1,990	1,590	1,325
10	14,400	7,200	4,800	3,600	2,880	2,400
$1/8$ unit vertical in 12 units horizontal (1-percent slope)						
3	960	480	320	240	192	160
4	2,040	1,020	681	510	408	340
5	3,520	1,760	1,172	880	704	587
6	5,440	2,720	1,815	1,360	1,085	905
7	7,800	3,900	2,600	1,950	1,560	1,300
8	11,200	5,600	3,740	2,800	2,240	1,870
10	20,400	10,200	6,800	5,100	4,080	3,400
$1/4$ unit vertical in 12 units horizontal (2-percent slope)						
3	1,360	680	454	340	272	226
4	2,880	1,440	960	720	576	480
5	5,000	2,500	1,668	1,250	1,000	834
6	7,680	3,840	2,560	1,920	1,536	1,280
7	11,040	5,520	3,860	2,760	2,205	1,840
8	15,920	7,960	5,310	3,980	3,180	2,655
10	28,800	14,400	9,600	7,200	5,750	4,800
$1/2$ unit vertical in 12 units horizontal (4-percent slope)						
3	1,920	960	640	480	384	320
4	4,080	2,040	1,360	1,020	816	680
5	7,080	3,540	2,360	1,770	1,415	1,180
6	11,080	5,540	3,695	2,770	2,220	1,850
7	15,600	7,800	5,200	3,900	3,120	2,600
8	22,400	11,200	7,460	5,600	4,480	3,730
10	40,000	20,000	13,330	10,000	8,000	6,660

For SI: 1 inch = 25.4 mm, 1 square foot = 0.0929 m².

Step two is to take that number and decide how many drains are to be used. Step three is to take the number of drains used and divide it into the maximum total rainfall on that surface. This is when Tables 1106.2 and 1106.3 are used. The total as calculated above is then used when entering both tables to determine the minimum pipe sizing required. Greater vertical sloping increases the volume potential and the appropriate $1/16$, $1/8$, $1/4$ slope tables should be used in Table 1106.3.

The rain leader or vertical piping can carry more water than the horizontal piping. Precautions should be taken to size the horizontal storm drain accordingly. Where horizontal drains intersect with others, make sure that Table 1106.3 is used to re-calculate the sizing.

Public Works Departments in your local jurisdiction will provide the information necessary for the appropriate connection to the public storm sewer system.

Scuppers and Leaders

Sizing of roof scuppers shall be in accordance with Table 1106.6 using the same sizing calculation. Both overflow drains and overflow scupper locations shall comply with the building code and are required to be in adjacent walls with the inlet rim 2 inches above the adjacent roof surface at the roof drains. Sometimes scuppers serve as both roof drains and overflow channels. The size would be equivalent to the two cross-sectional areas added together. Remember that this minimum area is net area so an allowance should be made for roofing or flashing to enter the scupper.

When using scuppers there should be a leader on the exterior side of the parapet leading to a

Storm Drainage

downspout that is sized to match the roof drain size. There should be a slot cut out of the exterior side 2 inches above the bottom of the leader to allow for overflow.

Testing is required in accordance with Section 312.8, which refers to a 5-pound air test or a water test with a 10-pound head of water. Backwater valves installed in a storm water system shall conform to Section 715, where it is required that plumbing fixtures (roof, deck or area drains) cannot be below the next upstream manhole in the public storm sewer. The rest of that section applies only if a backwater valve is required.

Storm water shall not be discharged into sewers designed and intended for sewage only. Conductor pipes shall not be used as either soil, waste or vent pipes, and soil, waste or vent pipes shall not be used as conductors. Floor drains shall not be connected to a storm drain.

Combined Storm and Sewer Systems

Although it has generally lost favor in recent years, where combined storm and sewer systems are permitted, the building storm drain shall be connected on the horizontal through a single wye fitting to the combined sewer at least 10 feet downstream from the soil stack.

Leaders and storm drains connected to a combined sewer shall be trapped. Individual storm-water traps shall be installed on the storm-water branches, or a single trap shall be installed on the main storm drain just before its connection with the building or public sewer. Storm-water traps shall be of the same material as the piping used and shall be the same size as the horizontal drain to which they are connected. A cleanout shall be installed on the building side of the trap.

Sizing of the combined sewer shall be accomplished using the method in Section 1106.3, which is based on that same table. Section 1108.1 goes into some detail about calculating a combined sanitary and storm sewer using two methods of identification.

Below-grade Storm Drain Systems

Underground storm drainage located inside the building must conform with material standards listed in Table 702.2 and storm-drainage systems located outside the building must conform to the material standards in Table 1102.4 with fittings conforming with the material standards in Table 1102.7.

Subsoil drains shall comply with the material standards listed in Table 1102.5. Subsoil drains shall be open-jointed, horizontally split or perforated pipe, and not less than 4 inches in diameter. If the building is subject to backwater, a sump, trapped area drain, dry well or other area approved for ponding has to be provided above ground. Subsoil sumps are not required to be gas tight or to have a vent.

Building subdrains located below the public sewer level shall discharge into a sump or holding tank, the contents of which are then pumped into the drainage system. Sump pumps, pits and discharge piping shall conform to Section 1113. The sump pump shall have adequate head capacity and the sump pit shall be not less than 18 inches in diameter with approved casing material and permanent flooring to support the pump. Discharge piping shall meet the requirements of Section 1102.2, 1102.3 or 1102.4 and shall include a gate valve and a full-flow check valve. Piping size and fittings shall be of the same size or larger than the pump discharge tapping.

TABLE 702.2
UNDERGROUND BUILDING DRAINAGE AND VENT PIPE

MATERIAL	STANDARD
Acrylonitrile butadiene styrene (ABS) plastic pipe	ASTM D 2661; ASTM F 628; CSA B181.1
Asbestos-cement pipe	ASTM C 428
Cast-iron pipe	ASTM A 74; CISPI 301; ASTM A 888
Coextruded composite ABS DWV sch 40 IPS pipe (solid)	ASTM F 1488
Coextruded composite ABS DWV sch 40 IPS pipe (cellular core)	ASTM F 1488
Coextruded composite PVC DWV sch 40 IPS pipe (solid)	ASTM F 1488
Coextruded composite PVC DWV sch 40 IPS pipe (cellular core)	ASTM F 1488
Coextruded composite PVC IPS - DR, PS140, PS200 DWV	ASTM F 1488
Copper or copper alloy tubing (Type K, L, M or DWV)	ASTM B 75; ASTM B 88; ASTM B 251; ASTM B 306
Polyolefin pipe	CSA CAN/CSA-B181.3
Polyvinyl chloride (PVC) plastic pipe (Type DWV)	ASTM D 2665; ASTM D 2949; ASTM F 891; CSA CAN/CSA-B181.2
Stainless steel drainage systems, Type 316L	ASME/ANSI A112.3.1

TABLE 1102.4
BUILDING STORM SEWER PIPE

MATERIAL	STANDARD
Acrylonitrile butadiene styrene (ABS) plastic pipe	ASTM D 2661; ASTM D 2751; ASTM F 628
Asbestos-cement pipe	ASTM C 428
Cast-iron pipe	ASTM A 74; ASTM A 888; CISPI 301
Concrete pipe	ASTM C 14; ASTM C 76; CSA A257.1; CSA CAN/CSA A257.2
Copper or copper-alloy tubing (Type K, L, M or DWV)	ASTM B 75; ASTM B 88; ASTM B 251; ASTM B 306
Polyvinyl chloride (PVC) plastic pipe (Type DWV, SDR26, SDR35, SDR41, PS50 or PS100)	ASTM D 2665; ASTM D 3034; ASTM F 891; CSA-B182.2; CSA CAN/CSA-B182.4
Vitrified clay pipe	ASTM C 4; ASTM C 700
Stainless steel drainage systems, Type 316L	ASME A112.3.1

TABLE 1102.7
PIPE FITTINGS

MATERIAL	STANDARD
Acrylonitrile butadiene styrene (ABS) plastic	ASTM D 2468; ASTM D 2661
Cast iron	ASME B16.4; ASME B16.12; ASTM A 888; CISPI 301; ASTM A 74
Chlorinated polyvinyl chloride (CPVC) plastic	ASTM F 437; ASTM F 438; ASTM F 439
Copper or copper alloy	ASME B16.15; ASME B16.18; ASME B16.22; ASME B16.23; ASME B16.26; ASME B16.29; ASME B16.32
Gray iron and ductile iron	AWWA C110
Malleable iron	ASME B16.3
Plastic, general	ASTM F 409
Polyethylene (PE) plastic	ASTM D 2609
Polyvinyl chloride (PVC) plastic	ASTM D 2464; ASTM D 2466; ASTM D 2467; CSA CAN/CSA-B137.2; ASTM D 2665
Steel	ASME B16.9; ASME B16.11; ASME B16.28
Stainless steel drainage systems, Type 316L	ASME A112.3.1

Engineer-designed Systems

Since roof drainage is part of roof design, it is a given that the design of the system will be part of the design of the building. The roof must be designed first and foremost to accommodate the weight of the maximum accumulated water. The placement of the drains and the heights of overflow drains, scuppers, or parapets will determine this maximum accumulation.

If the roof structure is engineered for controlled flow, a design must be submitted and approved based on the requirements in Section 105.4. Rainfall amounts will be based on Section 1106.1. Control devices shall be installed so that the rate of discharge does not exceed the values shown in Section 1109.1. Strainers shall protect these control devices.

TABLE 1102.5
SUBSOIL DRAIN PIPE

MATERIAL	STANDARD
Asbestos-cement pipe	ASTM C 508
Cast-iron pipe	ASTM A 74; ASTM A 888; CISPI 301
Polyethylene (PE) plastic pipe	ASTM F 405
Polyvinyl chloride (PVC) plastic pipe (type sewer pipe, PS25, PS50 or PS100)	ASTM D 2729; ASTM F 891; CSA-B182.2; CSA CAN/CSA-B182.4
Vitrified clay pipe	ASTM C 4; ASTM C 700
Stainless steel drainage systems, Type 316L	ASME A112.3.1

However, the scupper proportions should be such that if there is a blockage, the water will not pond more than the roof was designed for. Water weighs 62.4-pounds/cubic foot. This means that if the minimum roof live load design was determined by the *International Building Code* to be, say, 20 pounds/square foot (see IBC Section 1607.11.2), this would only allow approximately 4 inches of ponding before the design would have to be revised. However, this is only a guideline number because on most roofs ponding of water can lead to a cascading failure. This means that as water accumulates, it causes the roof to deflect and that causes more ponding and so forth. (See IBC Section 1611 for details of ponding design.)

Suffice it to say, a scupper that is wider than it is high is preferable to one that is taller than it is wide.

Certain quantities of continuous flow in a continuous flow building storm-drain system can come from the following devices; pumps, ejectors, air conditioning units or roof-drain control devices. Each device shall have a maximum discharge rate of 96 square feet, based on a rainfall rate of 1 inch.

On controlled roof systems, there is a minimum of two roof drains on any roof area less than 10,000 square feet and a minimum of four roof drains on roof areas more than 10,000 square feet.

8

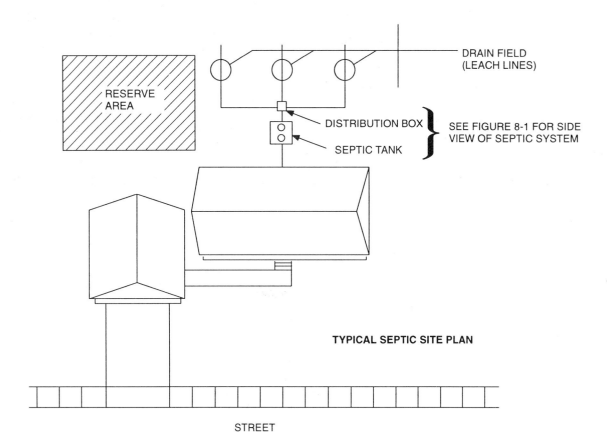

TYPICAL SEPTIC SITE PLAN

Private Disposal Systems

General

In previous chapters we have mentioned the term "septic systems" referred to in the IPSDC. In those brief and casual references, one hopes some interest was sparked concerning what to do in the absence of a connection to a public sewer. Taking the country as a whole (including the islands in the Pacific and Caribbean), and realizing that much of it is rural in nature, this question is considerably more than academic.

As discussed in Chapter 1, septic-type systems are not new. Over 4,000 years have passed and the concept behind effluent infiltrating the earth has not changed. In fact, this is still the end result of any private disposal system. Nonetheless, we seem to have quite a number of rules to follow for such a seemingly low-tech solution.

First things first. Let's begin with an overview of the standard septic design, starting with the connection from the building drain to the septic tank. From the tank, where the solids settle, the effluent is sent into the absorption system, where it percolates into the surrounding soil.

The land in which the absorption system is placed shall not have a slope greater than 20 percent. Careful consideration is given regarding the closeness to water supplies and waterways. The groundwater table can adversely affect the septic location. The absorption system shall not be placed in a fill area without approval. Roads, lot lines and structures have minimum separation distances required. Table 406.1 of the IPSDC conveniently lists these minimum horizontal distances.

After the site has been selected, soil borings are taken and the ground is evaluated for permeability. The test performed is called a percolation test.

Usually six borings are taken: three for the absorption area and three for the eventual replacement area (to be only used, one prays, many years later). In each of the borings, sand or gravel is placed at the bottom, a standard measure of water is poured in, and the rate of absorption is measured over time. Reports of the percolation test are then submitted to the code official, who may consult health department officials for advice. The report is then evaluated and a determination is made regarding the type of system(s) allowed.

Once permission is given on the design, the tank is connected to the building drain and buried in place (Note: septic tanks must meet the design standards shown in Chapter 5 of the IPSDC.) The outlet side of the septic tank is connected to the absorption system by means of perforated pipe facing downward on top of 6 inches of aggregate ranging from $1/2$ inch to $2^1/_2$ inches in size.

Aggregate, two inches deep, is then placed over the top of the perforated pipe and 9 inches of synthetic material or marsh hay is added on top of that. A minimum of 18 inches of soil backfill is then placed to provide the covering.

Other types of private sewage-disposal systems include pressure-distribution systems, mound systems, cesspools (usually temporary), and holding tanks for retaining a 5-day-minimum capacity when other systems are not available.

Details about private disposal systems can be researched further by obtaining a copy of the IPSDC.

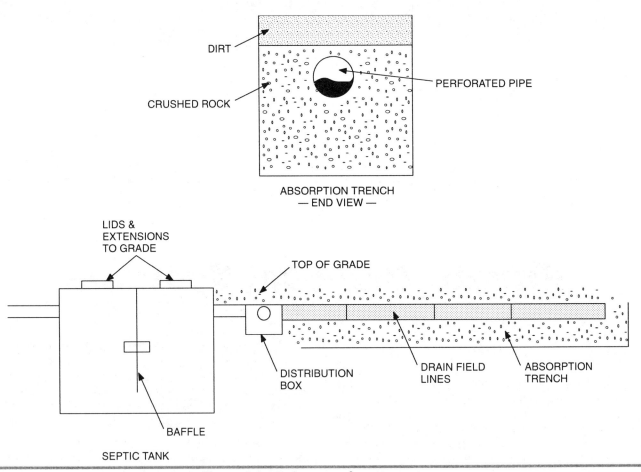

Figure 8-1 *Side view of septic system*

APPENDIX A

Comparison & Cross Reference: 1997 UPC and 2000 IPC

Introduction

"No Code is perfect. Nor is it likely that there will ever be a perfect code."

So begins the foreword to the publication *1997 IPC/1997 UPC: An Overview and Analysis* (ICBO, 1998). This pamphlet was written in cooperation with Arthur J. Lettenmaier, J.D., who has had extensive experience in the plumbing field (including serving as a licensed plumber and plumbing supervisor, plumbing plans examiner and code instructor). He was also formerly Director of Education and Publications and Executive Director for the International Association of Plumbing and Mechanical Officials (IAPMO, publisher of the UPC). This publication provided a comprehensive comparison between the two major plumbing codes in use in the United States since 1995. This foreword succinctly continues and puts the two codes into context as follows:

"It is difficult to develop a well-written code. It is an enormous challenge to put together language that allows everything that should be allowed and prohibits everything that should be prohibited. Science and technology are always advancing, and today, faster than ever. This compounds the problem of writing a comprehensive code because of the difficulty in anticipating new methods, materials and technologies. Nonetheless, the IPC and the UPC both take on this challenge. Both do an admirable job. Both codes are comprehensive. Both provide detailed requirements for conventional plumbing systems. Both contemplate new methods, materials and technologies.

"Neither book is perfect, but there are a few factors that make the IPC preferable as a code. Every section of the IPC has a descriptive title, not so with the UPC. The IPC is also better organized in that regulations are most likely to be found in the appropriate chapter. The UPC has duplicative requirements, some conflicting requirements, some misleading titles and some regulations in unexpected locations.

"Both codes give the code official discretion in certain instances. Some will be surprised to find that the UPC more frequently fails to provide specific guidance than does the IPC. The phrase, '*shall be approved by the Administrative Authority,*' is overworked in the UPC. To be of the greatest use, the text of the code should allow the installer, the inspector and the code official to distinguish between means of installation that will be approved and those that will not be approved."

If this is true, why do some claim that the IPC is technically less "stringent" than the UPC?

Source of Assumptions about the IPC

The venting provisions of the IPC represent practices used or that could be used in any part of this country or its territories. Some of them will look "strange" to installers, depending on where they come from or live. This often raises understandable but incorrect assumptions that:

- The IPC is deficient, putting people at risk from a health/safety standpoint.
- To adopt the IPC will totally change the way plumbing is designed, thus you will have to be *completely* retrained if you are going to maintain your involvement in the field of plumbing.

There are a number of published documents that detail some of the technical differences between these codes. They can be found in reference 2 cited in Chapter 1 of this guide. One goal for this appendix is to examine some differences between the IPC and UPC in order to show that the reactions embodied in the two assumptions above are not based on scientific or rational data. We will also go over the differences between the two codes chapter by chapter. But please note that there are many more similarities than differences.

Detailed Analysis

Let's first study the contention that the IPC provides for somehow deficient installation practices that result in unsafe health risks. The IPC evolved from three other existing plumbing codes. Two of these codes [*Standard Plumbing Code*™ (SPC) and *National Plumbing Code*™ (NPC)] contain venting methods similar to and/or identical to those published in the IPC. If these provisions worked improperly or in an unsafe manner, surely we would have been hearing about it by now from those areas of the United States where they are in use—the South, Northeast and into the Midwest. This represents at least 40 percent of the U.S. population. To say that the IPC is deficient or unsafe is to say that these other codes were deficient or unsafe, and this is just not the case. (For further details on how the IPC was developed, see Chapter 1 of this guide.)

What about the second statement that adopting the IPC will completely change the way one designs and installs plumbing, resulting in extensive retraining and so forth? Let's examine four areas where the IPC and the UPC differ substantially and that appear to be the practices used to support this statement.

Water Distribution System Sizing. This is one of the main areas where the basic difference between the IPC and the UPC methodology has been described as "performance-based" (IPC) versus "prescriptive-based" (UPC). A fair and technical in-depth study of both codes reveals that each contains a measure of both performance and prescriptive provisions. Jurisdictions that adopt and administer codes with performance-based sizing considerations design their certification training accordingly.

Does this mean that the prescriptive methods found in the UPC are invalid? By no means! The prescriptive method is based on "performance," albeit under conservative assumptions of what a "typical" installation will entail. A point of interest is that Table 6-5 of the UPC is based on a velocity of 10 feet per second (see *Official Magazine*, May/June 1992, page 16, published by IAPMO). Section 610.12 of the UPC restricts the velocity for copper tubing to 8 feet per second for cold water and 5 feet per second for hot water. Yet Section 610.1 exempts those systems sized per Table 6-5. This illustrates the difficulty of designing a "one size fits all" prescriptive sizing chart.

In fact, within the UPC itself you have a choice of the prescriptive method (Chapter 6) or the performance-based method (Appendix A, Appendix E or Appendix L). To be proficient in plumbing design, one needs to understand all these methods.

So, will training really change drastically if there no longer are prescriptive tables such as are found in UPC Chapter 6? No! Must you abandon the prescriptive method should the IPC be adopted locally? Of course not! Conservative and prescriptive methods can meet the minimum provisions of the IPC.

In fact, in recognition of the fact that some jurisdictions and plumbers wanted to retain their prescriptive methods, the IPC code development process is currently reviewing some prescriptive sizing criteria for inclusion in the 2003 IPC.

Sanitary Drain Sizing Methods. In this area both codes utilize prescriptive tables. The tables differ somewhat because the IPC allows for varying degrees of slope and also gives consideration to the number of branch intervals that are connecting with the stack. In other words, it allows more variables than the UPC tables do. One could say that the UPC has one-size-fits-all tables, while the IPC gives more choices.

While the tables in the two codes are different, they are still just "tables" that were developed using different criteria. It really is no harder to read one table than another except that it is human nature to look askance at something new. When one realizes that as technology advances, the older ways have to be modified, then it should not be assumed that any change is somehow "unsafe."

In today's world of performance-based construction, the use of tables is commonplace on construction plans. Be it shear-nailing patterns, anchoring and hold-down sizes/details for seismic bracing, or whatever, each designer has a different way of showing the information on their set of construction plans. A table is only a tool with which to present concise information. It has a technical basis founded on research, and the user merely has to read and apply the information therein. Actually, the user should find that the tables in the IPC are as simple to use as are those found in the UPC. With very few exceptions, a drainage system sized per the UPC will meet the minimum provisions of the IPC and vice-versa.

Height Limitation of ABS/PVC in DWV Systems. The limitation of ABS/PVC in DWV systems to three floors above grade was only found in the UPC, not the IPC or in the documents upon which it was based. In reality, the reason given for this prohibition—protection of people from smoke caused by fires—is a Building Code issue, not a Plumbing Code issue. The Building Code has numerous safeguards that mitigate any supposed fire hazard, including through-penetration fire-stopping for pipes that pass through fire-resistive assemblies or membranes. Interestingly, it is metallic piping spreading fire through conduction of heat while plastic pipes merely melt. The fire-stopping material then expands into the hole left by the melted plastic.

The fact that the IPC and other plumbing codes never contained this limitation, and that most jurisdictions adopting the UPC had to amend the code to strike this provision, proves that such a provision really had no scientifically valid basis for existence. (Note that the 2000 UPC has amended this provision, but not completely enough to cause the UPC to defer to the Building Code, as does the IPC.)

Venting Methods. While many will claim that in this aspect the IPC and the UPC are "miles apart," they really are not as far apart as one might initially think. Both codes contain a "conventional" sizing method, (another "one-size-fits-all" method), and while the UPC contains four additional specific venting methods, the IPC contains nine!

All these conventional methods are quite similar to each other, except that the UPC requires what is termed "cross-sectional area venting," and for a water closet, the minimum vent size is arbitrarily set at 2 inches in diameter. Are these provisions absolutely necessary? Apparently not, given the fact that jurisdictions that do *not* administer the UPC are not experiencing failures of their conventional plumbing systems!

Is It Too Costly to Change Codes?

Would changing from the UPC to the IPC pose a major impact? Not when you consider that the conventional venting provisions of the UPC will meet or frequently exceed the *minimum* national-consensus provisions contained in the IPC. And what about the other specific venting applications of the IPC? *They are options* and are not mandatory. It's the choice of the designer or installer in consultation with and agreement by the owner that determines their use.

Can the plumbing industry in local jurisdictions handle such changes when they shift from the UPC to the IPC? The State of Washington went from the 1991 to the 1997 UPC in July 1998 and were confronted with:

- More than 160 changes from the 1991 edition to the new edition.
- More than 80 amalgamated portions that had been added to make the new edition without written commentary explaining their application.
- Two completely new chapters.
- Five new appendix chapters.
- Complete reformatting of the entire code.
- More than 60 state amendments added.

Did the plumbing industry in Washington fold up and die? Not at all! The industry and local jurisdictions accepted the changes, began training accordingly and carried on as they did with every code cycle change in the past. And so it will be if any jurisdiction adopts the IPC over any previously adopted code (except that there will likely be *much less need* for state amendments with the IPC). Life goes on.

General Differences Between the 2000 IPC and the 1997 UPC

What follows are differences between the IPC and UPC (note that we are showing the "differences." If we gave both differences and similarities, the list would be at least three times as long):

Chapter 1 Administration
- The IPC contains language that more clearly defines the requirements for plan preparation and review, similar to the IBC.

- The UPC allows jurisdictional authority to grant deviations from code provisions in existing construction.
- The IPC allows an emergency stop-work order to be given verbally and also requires a "work resumption" statement to be given.
- The IPC requires plans showing piping penetrations to include materials and methods for maintaining structural safety, fire-resistive rating and fire blocking for buildings more than two stories in height.
- The UPC requires the permit revocation to be made in writing.
- The UPC allows a plumbing system to be put into use without inspection, provided it was installed to replace an existing system in an occupied space, and that the request for inspection was filed not more than 72 hours after completion of the work. No portion of the installation shall be concealed prior to inspection.
- The UPC requires testing of plumbing systems in moved buildings.
- The IPC allows a "temporary connection" to be made for a temporary certificate of occupancy.
- The IPC contains language that places restrictions on employees of the "plumbing department" that might be deemed a conflict of interest. The UPC does not address this issue.
- The IPC requires that inspectors carry proper identification when performing their duties.
- The IPC contains language regarding special inspections or evaluation services that is not found in the UPC.
- The IPC outlines an appeal process, including the design of the appeals board and its realm of authority. The UPC does not include "appeal" language.

Chapter 2 Definitions
- While the definition sections in the two code documents vary, the vast majority of differences are due to particular items or language peculiar to each code. For instance, what the UPC defines as a "wet vent," the IPC defines as a "stack vent."
- Another difference is where the codes use different terms to identify the same thing. For instance, what the IPC defines as the "Code Official" is defined in the UPC as the "Administrative Authority."
- One must also recognize that some items are defined in one code and not in the other. One such item is "air-admittance valve" in the IPC, which is not found in the UPC. This may be a result of unique items identified and/or allowed in a particular code document, or it may just be the authors' discretion in determining which items were deemed necessary to define.

Chapter 3 General Regulations
- The IPC contains more detailed language regarding testing criteria, similar to the IBC.
- The UPC requires sewage disposal systems to be located on the same lot as the structure and prohibits the sale or transfer of property that would restrict access to said facilities.
- The UPC contains specific wording that does not allow concealment of cracks, holes or other defects or imperfections by means of paint, wax, tar or other leak-sealing repair agents.
- The UPC prohibits the use of certain manufactured drainage fittings that are not specifically restricted in the IPC. (See UPC 311.0 and IPC 605.9, 707.1, Table 706.3.)
- The UPC requires joints of dissimilar metals to be exposed (except for necessary valves) regardless of manufacturer's specifications.
- The IPC requires plumbing installations to be recessed in walls or otherwise protected when located in parking garages or along alleys and driveways.
- The UPC restricts the use of counter-flashing material that obstructs the cross-sectional area of the vent opening.
- The codes are different in their requirements for nail plates. (See UPC 313.9 and IPC 305.8.) The UPC requires sleeves, in general, to have a $1/2$-inch clearance around the pipe and to be sealed watertight. The IPC requires the sleeve to be two pipe sizes larger than the penetrating pipe when passing through a foundation wall or under a footing.
- The IPC requires seismic support; sway bracing on 4-inch or larger pipe at every change of direction of 45° or more, and at changes in pipe diameter greater than two pipe sizes; specific support at bases of stacks with concrete or mortar materials or metal brackets. The UPC requires horizontal cast-iron hubless piping in excess of 4-foot lengths to be supported within 18 inches of the coupling on each side.
- Support of piping varies to some degree. (See UPC Table 3-1 and IPC Table 308.5.)
- The UPC contains specific requirements for food-handling establishments. The IPC contains no such section.
- The UPC specifies types of gauges to be used for various types of pressure tests.
- The IPC contains a section regulating plumbing that may be located in a flood zone. The IPC outlines requirements for light, ventilation, wall finish and compartment design for washrooms and toilet rooms.

Chapter 4 Fixtures, Faucets and Fixture Fittings
- The IPC allows urinals with invisible seals to be installed for private use, whereas the UPC does not allow them under any condition.
- The UPC does not contain provisions regulating space requirements of lavatories or for water closet compartments. The IPC requires 21 inches in front of a water closet, compared to 24 inches in the UPC. The IPC requires a 15-inch clearance from a urinal to the side wall and 30 inches from center to center. The UPC requires a 12-inch clearance from a urinal to the side wall and 24 inches from center to center.
- The UPC allows more than one urinal to be supplied by an automatic flushing device. The IPC allows only one urinal per any flushing device.
- The UPC requires floor drains in certain toilet rooms, commercial kitchens and in laundry rooms of commercial buildings. The IPC requires them in coin-operated laundry facilities.
- The minimum square-inch requirement for shower compartments is 1,024 per the UPC, and 900 per the IPC.
- The UPC outlines distance requirements for drains located in gang showers.
- The UPC requires that the plumbing system (DWV and potable water) be installed to the point of the fixture connection where future plumbing is terminated at the location of the trap for future fixture.
- The IPC requires fixtures to be bolted through walls, and pipes or traps to be not exposed in mental health centers.
- The IPC allows bottled water coolers to substitute for drinking fountain requirements.
- The IPC contains language specifying wall and floor space material around a urinal installation.
- The IPC contains specific requirements for healthcare facilities not found in the UPC.

Chapter 5 Water Heaters
- The UPC defines a water heater based on the temperature of the water delivered, whereas the IPC defines it based on the use of the water delivered.
- The UPC includes a clothes closet as a prohibited location for water heaters and does not differentiate whether it is in, or accessed from, a sleeping room or bathroom.
- The IPC requires a drain pan equipped with a minimum $3/4$-inch drain where damage may occur due to leakage. The UPC also requires a drain pan with a minimum $3/4$-inch drain in attics or floor/ceiling assemblies.

Chapter 6 Water Supply and Distribution
- The IPC contains provisions regulating private wells that are not found in the UPC.
- The IPC does not state who is the party responsible for assuring periodic testing of backflow assemblies.
- The IPC allows single-wall heat exchangers when nontoxic fluids/gases are utilized. The UPC requires double-wall heat exchangers unless Appendix L is adopted locally.
- The UPC requires that hose bibbs subject to freezing must be listed self-draining types with integral backflow preventers.
- The UPC does not allow backflow assemblies to be installed in an area containing fumes that are toxic, poisonous or corrosive. The IPC refers to the standards that cover this aspect.
- The IPC requires that the water supply for all hospital fixtures be protected with a backflow preventer.
- The IPC allows plastic water-service material to terminate within a building. The UPC does not list ABS or polybutylene as approved materials. The IPC allows type M copper to be installed below ground inside a building.
- The IPC requires a number of shut-off valve locations that are not required in the UPC, such as at the base of a water riser or water downfeed in certain occupancies.
- The UPC contains criteria that allow and regulate mechanically formed tee fittings.
- The UPC requires a minimum residual pressure of 15 psi to each fixture, whereas the IPC may allow this to be as low as 4 psi, depending on the type of fixture.
- The UPC requires a minimum 12-inch depth below the frost line as opposed to 6 inches in the IPC.
- The UPC does not require any physical separation of water lines and waste lines in a trench when the waste-line material conforms to that required for inside a building.
- The UPC requires that copper joints in water lines below a slab floor be brazed.
- The UPC addresses location of water lines in relation to adjacent lots. That is not discussed in the IPC.
- The UPC allows site-built air chambers in lieu of manufactured/listed mechanical devices for the control of water hammer.
- The UPC provides prescriptive design criteria for sizing water-distribution systems, whereas the IPC provides performance criteria as a minimum that a design must conform to. (However, Appendix E of the IPC has provisions for recommended water-sizing procedural methods.) Chapter 29 of the *International Residential Code*® also provides prescriptive design

criteria tables for residential water supply systems.
- The IPC specifies valve identification for service and hose-bibb valves that are not adjacent to the fixture served.
- The IPC contains language regarding water tank overflows, wells and healthcare facilities that are not covered in the UPC.

Chapter 7 Sanitary Drainage
- The IPC provides sizing criteria based on grade-slope levels of $1/16$ inch per foot for 8-inch or larger diameter pipe. The sizing criteria differ slightly between codes.
- The UPC requires that double sanitary tees may be used only when the barrel of the fitting is two pipe sizes larger than the largest inlet.
- The UPC requires floor drains adjacent to certain restaurant fixtures and appliances that are not specifically required in the IPC.
- The IPC allows $1/4$ bends in some installations (2-inch and smaller) that require a longer sweep according to the UPC.
- The IPC does not require upper terminal cleanouts and does not grant the same exceptions to cleanouts that are allowed in the UPC.
- The minimum clearance required for cleanout access is less restrictive in the UPC, except that the IPC does not specify a minimum distance from an access door to a cleanout.
- The IPC does not require a dual-pump arrangement for sewage ejectors in a public use occupancy.
- The IPC does not contain requirements for suds relief. However, the IPC requires that connections to offsets and bases of stacks be 10 pipe sizes downstream of the offset or base of stack.
- The IPC contains provisions regulating offsets in buildings of five or more stories.
- The IPC contains special provisions for healthcare facilities and computerized drainage system design not found in the UPC.

Chapter 8 Indirect/Special Waste
- The minimum size requirements for indirect waste pipes differs slightly.
- The UPC restricts indirect waste pipes for sterilizers to no more than 15 feet.
- The UPC requires indirect waste pipes exceeding 5 feet but less than 15 feet to be directly trapped at the fixture connection with a vent being required for such in excess of 15 feet. The IPC requires a trap when the indirect waste pipe exceeds 2 feet in developed length horizontally or 4 feet in total developed length. No venting requirements are outlined in the IPC.
- The maximum height of a standpipe above its trap is 30 inches per the UPC, while the IPC permits standpipes up to 42 inches in height.
- The IPC does not contain any provision for drinking fountains to be installed with indirect wastes.

Chapter 9 Vents
- The UPC requires that the aggregate cross-sectional area of all vent pipes be equal to or greater than the cross-sectional area of the largest required building drain served. Any vent exceeding 40 feet, according to the IPC, must be increased one pipe size, while the UPC contains maximum length limitations per pipe size with a limit on horizontal installations unless the vent is increased one pipe size.
- The UPC does not require vents to be graded. It also does not allow side-inlet quarter bends or air-admittance valves.
- Extension through the roof is 6 inches in the UPC, but is not stated in the IPC. No provisions are found in the IPC pertaining to outdoor installation. Vent terminations above openings are required to be 3 feet per the UPC and 2 feet per the IPC. The criteria for frost protection is slightly different in each code.
- The various types of venting systems are different in design and name. (See diagrams for IPC venting installations in Chapter 9 of this guide.)

Chapter 10 Traps
- The IPC is less restrictive for lengths and grade of trap arms. (See Table 906.1.)
- The IPC allows the slip joint of a trap arm to serve as a fixture drain cleanout. (See Section 708.7, Exception.)

Chapter 11 Storm Drainage
- The UPC contains provisions for areaway drains, paved areas, and filling stations or vehicle washing establishments that are not specifically covered in the IPC.
- There are very slight differences in sizing criteria for secondary roof drainage.
- The UPC requires rainwater sumps serving public use occupancy buildings to be provided with dual pumps arranged to function alternately in case of overload or mechanical failure.
- The IPC addresses roof design for maximum ponding.

Chapter 12 Fuel Piping
- The IPC addresses fuel piping in Appendix G, which differs significantly from the UPC provisions of Chapter 12. There is a new ICC promulgated document called the *International Fuel Gas Code®*.

Conclusion

The bottom line is that while the IPC and the UPC are technically different, they really are not substantially as different as they may initially appear. This is especially so when one considers the minimal impact that would occur should your jurisdiction adopt the IPC instead of the UPC. There would be even less impact in jurisdictions that are now using the other plumbing codes that provided the basis of and are similar to the IPC. For the most part, as stated earlier, one can install a conventional plumbing system in conformance with the UPC and still easily meet the minimum provisions of the IPC.

Since this doesn't work the other way around, it shows that the UPC is conservative and is not taking advantage of labor and material savings allowed by modern plumbing technology. (Appendix B of this document shows how the IPC designs can be cost effective for residential buildings.) Systems designed and built to either code are safe and effective but the IPC is based more on engineering principles and offers more choices.

Reference:

1. *1997 IPC/1997 UPC: An Overview and Analysis.* ICBO and Professional Plumbing Seminars, February 1998.
2. UPC-IPC (1997-2000): Comparison & Cross Reference.

(Please see page 95 for UPC-IPC Cross Reference)

Comparison & Cross Reference: 1997 UPC and 2000 IPC

| 1997 Uniform Plumbing Code | 2000 International Plumbing Code |

Chapter 1: Administration

1997 Uniform Plumbing Code	2000 International Plumbing Code
101.0 Title, Scope and General	
101.1 Title	101.1 Title
101.2 Purpose	101.3 Intent
101.3 Plans Required	106.5.2 Validity
101.4 Scope	101.2 Scope
101.4.1	102.2 Existing installations
101.4.1.1 Repairs and Alterations	102.4 Additions, alterations or repairs
101.4.1.1.1	
101.4.1.1.2	703.4 Existing building sewers and drains
101.4.1.1.3	102.2 Existing installation 701.2 Sewer required
101.4.1.2 Maintenance	102.3 Maintenance
101.4.1.3 Existing Construction	701.2 Sewer required
101.4.2	102.4 Additions, alterations or repairs
101.5 Application to Existing Plumbing System	
101.5.1 Additions, Alterations or Repairs	102.4 Additions, alterations or repairs
101.5.2 Health and Safety	102.9 Requirements not covered by code 108.6 Abatement of violation 108.7 Unsafe plumbing 108.7.3 Connection after order to disconnect
101.5.3 Existing Installation	102.2 Existing installations
101.5.4 Changes In Building Occupancy	102.5 Change in occupancy
101.5.5 Maintenance	102.3 Maintenance
101.5.6 Moved Buildings	102.7 Moved buildings
102.0 Organization and Enforcement	
102.1 Administrative Authority	103.2 Appointment
102.2 Duties and Powers of the Administrative Authority	**104 DUTIES & POWERS OF THE CODE OFFICIAL** 104.1 General

1997 Uniform Plumbing Code	2000 International Plumbing Code
102.2.1	103.3 Deputies
102.2.2 Right of Entry	104.5 Right of entry 104.6 Identification
102.2.3 Stop Orders	108.5 Stop work orders
102.2.4 Authority to Disconnect Utilities in Emergencies	108.6 Abatement of violation 108.7.2 Authority to disconnect service utilities 108.7.3 Connection after order to disconnect
102.2.5 Authority to Condemn	108.6 Abatement of violation 108.7.1 Authority to condemn equipment 108.7.3 Connection after order to disconnect
102.2.6 Liability	103.5 Liability
102.3 Violations and Penalties	
102.3.1 Violations	108.1 Unlawful acts
102.3.2 Penalties	108.4 Violation penalties 108.5 Stop work orders
103.0 Permits and Inspections	
103.1 Permits	
103.1.1 Permits Required	106.1 When required
103.1.2 Exempt Work	106.2 Exempt work
103.1.3 Licensing	106.4 By whom application is made
103.2 Application For Permit	
103.2.1 Application	106.3 Applications for permit
103.2.1.5	106.4 By whom application is made
103.2.2 Plans and Specifications	106.3.1 Construction documents
103.2.3 Information on Plans and Specifications	106.3.1 Construction documents
103.3 Permit Issuance	106 Permits
103.3.1 Issuance	106.5 Permit issuance 106.5.1 Approved construction documents 106.6 Fees
103.3.2 Retention of Plans	104.8 Department records 106.5.6 Retention of construction documents
103.3.3 Validity of Permit	106.5.2 Validity

1997 Uniform Plumbing Code	2000 International Plumbing Code
103.3.4 Expiration	106.5.3 Expiration 106.5.4 Extensions
103.3.5 Suspension or Revocation	106.5.5 Suspension or revocation of permit
103.4 Fees	
103.4.1 Permit Fees	106.6 Fees 106.6.2 Fee schedule (Also see **Appendix A**)
103.4.4 Investigation Fees: Work Without a Permit	106.6.1 Work commencing before permit issuance
103.4.5 Fee Refunds	106.6.3 Fee refunds
103.5 Inspections	**107 INSPECTION AND TESTING**
103.5.1 General	104.4 Inspections 107.1 Required inspections and testing
103.5.1.1 Inspection	107.1 Required inspections and testing
103.5.1.2 Scope	107.1 Required inspections and testing
103.5.1.3 Covering or Using	107.1 Required inspections and testing
103.5.3 Testing of Systems	**107 INSPECTION AND TESTING** 107.3 Testing 107.3.1 New, altered, extended or repaired systems 601.4 Tests 901.5 Tests
103.5.3.2 Test Waived	107.3.1 (2.) New, altered, extended or repaired systems
103.5.4 Inspection Requests	**108 VIOLATIONS**
103.5.4.1 Advance Notice	312.1 Required tests
103.5.4.2 Responsibility	107.3.2 Equipment, material and labor for tests, 312.1 Required tests
103.5.5 Other Inspections	104.4 Inspections 105.3 Required testing
103.5.5.1 Defective Systems	104.4 Inspections 105.3 Required testing
103.5.5.2 Moved Structures	102.7 Moved buildings
103.5.6 Reinspections	102.3 Maintenance 107.3.3 Reinspection and testing

1997 Uniform Plumbing Code	2000 International Plumbing Code
103.5.6.1 Corrections	107.1 Required inspections and testing 104.7 Notices and orders 108.2 Notice of violation
103.5.6.2 Retesting	107.3.3 Reinspection and testing
103.5.6.3 Approval	**105 APPROVAL** 107.1 Required inspections and testing 107.5 Approval
103.6 Connection Approval	108.7.3 Connection after order to disconnect
103.6.1 Energy Connections	108.7.3 Connection after order to disconnect
103.6.2 Other Connections	108.7.3 Connection after order to disconnect
103.6.3 Temporary Connections	107.6 Temporary connection
103.7 Unconstitutionality	101.4 Severability
103.8 Validity	101.4 Severability
Table 1-1	**Appendix A**

Chapter 2: Definitions

201.0 General	201.1 Scope 201.4 Terms not defined
202.0 Definition of Terms	**202 GENERAL DEFINITIONS**

Chapter 3: General Regulations

301.0 Materials - Standards and Alternates	
301.1 Minimum Standards	102.8 Referenced Codes & Standards 303.2 Installation of materials See Chapter 4 for individual fixture approval
301.1.1 Approvals	303.4 Third-party testing and certification
301.1.2 Marking	303.1 Identification
301.1.3 Standards	301.5 Pipe, tube and fitting sizes 303.2 Installation of materials
301.1.4 Existing Buildings	102.4 Additions, alterations or repairs 105.1 Modifications

1997 Uniform Plumbing Code	2000 International Plumbing Code
301.2 Alternate Materials and Methods	105.2 Alternative materials, methods and equipment 105.4 Alternative engineered design 701.8 Engineered systems
301.2.1 Intent	105.2 Alternative materials, methods and equipment
301.2.2 Compliance	105.4.1 Design criteria
301.2.3 Requirements	105.2 Alternative materials, methods and equipment,
301.2.4 Testing	105.3 Required testing 105.3.2 Testing agency 107.1.1 Approved agencies
301.2.5	105.3 Required testing 105.3.1 Test methods
301.2.6	105.3 Required testing
302.0 Iron Pipe Size (I.P.S.) Pipe	
303.0 Disposal of Liquid Waste	701.4 Sewage treatment
304.0 Connections to Plumbing System Required	301.3 Connections to drainage systems
305.0 Sewers Required	701.2 Sewer required
305.1	
305.2	
305.3	
306.0 Damage to Drainage System or Public Sewer	
306.1	302.1 Detrimental or dangerous materials
306.2	
307.0 Industrial Wastes	302.2 Industrial wastes 701.5 Damage to drainage system or public sewer
307.1	302.2 Industrial wastes
307.2	701.4 Sewage treatment
308.0 Location	
308.1	
308.2	

1997 Uniform Plumbing Code	2000 International Plumbing Code
309.0 Improper Location	301.6 Prohibited locations 310.2 Location of fixtures and piping
310.0 Workmanship	
310.1	
310.2	
310.3	
310.4 Installation Practices	303.2 Installation of materials
311.0 Prohibited Fittings and Practices	707.1 Prohibited joints **910 WASTE STACK VENT**
311.1	
311.2	
311.3	
311.4	910.1 Waste stack vent permitted
311.5	704.2 Change in size 706.2 Obstructions 1101.6 Fittings and connections
311.6	
311.7	
311.8 Screwed Fittings	Threaded joints: 605.10.3, 605.12.3, 605.14.4, 605.16.3, 605.18.1, 605.21.3, 705.2.3, 705.4.3, 705.9.4, 705.12.1, 705.14.3
312.0 Independent Systems	701.3 Separate sewer connection
313.0 Protection of Piping, Materials and Structures	**305 PROTECTION OF PIPES AND PLUMBING SYSTEM COMPONENTS**
313.1	305.1 Corrosion 305.2 Breakage 305.5 Pipes through or under footings or foundation walls 308.8 Expansion joint fittings
313.2	301.2 System installation 305.1 Corrosion 305.3 Stress and strain 308.8 Expansion joint fittings
313.3	301.2 System installation 307.4 Trench location

1997 Uniform Plumbing Code	2000 International Plumbing Code
313.4	305.6.1 Sewer depth
313.5	305.1 Corrosion 305.8 Protection against physical damage 305.9 Protection of components of plumbing system
313.6	305.6 Freezing
313.7	301.2 System installation 307.3 Penetrations of floor-ceiling assemblies and fire-resistance-rated assemblies
313.8 Waterproofing of Openings	301.2 System installation 305.7 Waterproofing of openings
313.9	305.8 Protection against physical damage
313.10 Sleeves	305.4 Sleeves 305.5 Pipes through or under footings or foundation walls
313.10.1	305.2 Breakage
313.10.2	305.5 Pipes through or under footings or foundation walls
313.10.3	305.5 Pipes through or under footings or foundation walls
313.11	307.2 Cutting, notching or bored holes
313.12 Ratproofing	**304 RODENTPROOFING** 304.1 General
313.12.1	304.2 Strainer plates
313.12.2	304.3 Meter boxes
313.12.3	304.4 Openings for pipes
314.0 Hangers and Supports	**308 PIPING SUPPORT** 308.1 General
314.1	308.5 Interval of support
314.2	
314.3	306.1 Support of piping
314.4	308.3 Materials
314.5	308.4 Structural attachment

1997 Uniform Plumbing Code	2000 International Plumbing Code
314.6	308.3 Materials
314.7	
Table 3-1	308.3 Materials
314.8	
315.0 Trenching, Excavation and Backfill	**306 TRENCHING, EXCAVATION AND BACKFILL**
315.1	301.2 System installation 307.4 Trench location
315.2	306.4 Tunneling
315.3 Open Trenches	107.1(1) Required inspections and testing
315.4	306.2.1 Overexcavation 306.2.3 Soft loadbearing materials 306.3 Backfilling
316.0 Joints and Connections	
316.1 Type of Joints	
316.1.1 Threaded Joints	605.21.3 Threaded joints 705.2.3 Threaded joints 705.9.4 Threaded joints 705.14.3 Threaded joints
316.1.2 Wiped Joints	605.13 Gray iron and ductile iron joints 705.13.2 Wiped 705.16.5 Lead pipe to other piping material
Table 3-2	308.5 Interval of support 308.9 Stacks
Table 3-2, footnote 2	308.6 Sway bracing
316.1.3 Soldered Joints	705.9.3 Soldered joints 705.10.3 Soldered joints
316.1.4 Flexible Compression Factory-Fabricated Joints	Mechanical joints: 605.10.1, 605.12.2, 605.14.2, 605.15.3, 605.16.1, 605.17.2, 605.18.2, 605.19.3, 605.20.3, 605.21.1, 705.2.1, 705.4.2, 705.5.3, 705.7.1, 705.8.1, 705.9.2, 705.10.2, 705.12.2, 715.14.1
316.1.5 Solvent Cement Plastic Pipe Joints	705.14.2 Solvent cementing

1997 Uniform Plumbing Code	2000 International Plumbing Code
316.1.6 Brazing and Welding	605.6.1.2 Brazed joints 605.12.1 Brazed joints 605.12.4 Welded joints 705.4.1 Brazed joints 705.4.4 Welded joints 705.9.2 Mechanical joints 705.9.5 Welded joints 705.10.1 Brazed joints 705.12.1 Threaded joints
316.1.7 Pressure-Lock Type Connection	Mechanical joints: 605.10.1, 605.12.2, 605.14.2, 605.15.3, 605.16.1, 605.17.2, 605.18.2, 605.19.3, 605.20.3, 605.21.1, 705.2.1, 705.4.2, 705.5.3, 705.7.1, 705.8.1, 705.9.2, 705.10.2, 705.12.2, 715.14.1
316.2 Special Joints	605.22 Joints between different materials
316.2.1 Copper Tubing to Screw Pipe Joints	605.22.1 Copper or copper-alloy tubing to galvanized steel pipe 705.16.1 Copper or copper-alloy tubing to cast-iron hub pipe
316.2.3 Plastic Pipe to Other Materials	605.22.2 Plastic pipe or tubing to other piping material 705.16.4 Plastic pipe or tubing to other piping material
316.3 Flanged Fixture Connections	405.4 Floor and wall drainage connections
316.4 Prohibited Joints and Connections	706.2 Obstructions
316.4.1 Drainage System	706.2 Obstructions
316.4.2	420.4 Water closet connections 704.2 Change in size
317.0 Increasers and Reducers	
318.0 Food Handling Establishments	
319.0 Test Gauges	**312 TESTS AND INSPECTIONS**

1997 Uniform Plumbing Code	2000 International Plumbing Code
Chapter 4: Plumbing Fixtures and Fixture Fittings	
401.0 Materials – General Requirements	
401.1 Quality of Fixtures	402.1 Quality of fixtures 407.1 Approval 408.1 Approval 412.1 Approval 416.1 Approval 416.2 Cultured marble lavatories 417.1 Approval 418.1 Approval
401.2 Lead	402.4 Sheet lead
402.0 Water-Conserving Fixtures and Fittings	313.1 General 401.3 Water conservation
402.1	
402.2	
402.3 Water Closets	401.3 Water conservation
402.4 Urinals	401.3 Water conservation
402.5 Non-Metered Faucets	401.3 Water conservation
402.6 Metered Faucets	401.3 Water conservation
402.7 Kitchen Faucets	401.3 Water conservation
402.8 Shower Heads	401.3 Water conservation
402.9 Installation	303.2 Installation of materials
403.0 Materials – Alternates	
403.1 Zinc Alloy Components	303.2 Installation of materials
404.0 Overflows	405.7 Design of overflows
405.0 Strainers and Connections	
405.1 Strainers	416.3 Lavatory waste outlets 417.3 Shower waste outlet 418.2 Sink waste outlets
405.2 Connections	425.2 Access to concealed connections 705.17 Drainage slip joints

1997 Uniform Plumbing Code	2000 International Plumbing Code
405.3	407.2 Bathtub waste outlets 416.3 Lavatory waste outlets 418.2 Sink waste outlets
405.4	
406.0 Prohibited Fixtures	
406.1	401.2 Prohibited fixtures and connections 410.2 Prohibited location
406.2 Prohibited Urinals	401.2 Prohibited fixtures and connections 419.1 Approval
406.3	407.1 Approval
406.4 Concealed Fouling Surfaces	402.1 Quality of fixtures
407.0 Special Fixtures and Specialties	
407.1 Water Connections	423.1 Water connections
407.2	
407.3 Approved	402.2 Materials for specialty fixtures 423.2 Approval
408.0 Installation	
408.1 Cleaning	405.2 Access for cleaning
408.2 Joints	405.5 Water-tight joints
408.3 Securing Fixtures	405.4 Floor and wall drainage connections
408.4 Wall-Hung Fixtures	405.4.3 Securing wall-hung water closet bowls
408.5 Securing Floor-Mounted, Back-Outlet Water Closet Bowls	405.4 Floor and wall drainage connections
408.6 Setting	405.3 Setting 405.3.1 Water closets, lavatories and bidets 405.3.2 Urinals

Contractor's Guide to the Plumbing Code

1997 Uniform Plumbing Code	2000 International Plumbing Code
408.7 Installations for the Handicapped	404.1 Where required 404.1.1 Unisex toilet and bathing rooms 404.2 Unisex toilet and bathing rooms 404.2.1 Standard 404.2.2 Required fixtures 404.2.2.1 Unisex toilet rooms 404.2.2.2 Unisex bathing rooms 404.2.3 Location 404.2.3.1 Prohibited location 404.2.4 Clear floor space 404.2.5 Privacy 404.2.6 Signage
408.8 Supply Fittings	405.1 Water supply protection 608.2 Plumbing fixtures
409.0 Water Closets	
409.1	420.2 Water closets for public or employee toilet facilities
409.2 Water Closet Seats	420.3 Water closet seats
410.0 Urinals	608.2 Plumbing fixtures
411.0 Flushing Devices for Water Closets and Urinals	
411.1 Flushing Devices Required	425.1 Flushing devices required 425.1.1 Separate for each fixture
411.2 Automatic Flushing Tanks	425.1.1 Separate for each fixture
411.3 Flushometer Valves	425.1.1 Separate for each fixture
411.4 Water Supply for Flush Tanks	425.1 Flushing devices required 425.3 Flushometer valves and tanks 425.4 Flush tanks
411.5 Flush Valves in Flush Tanks	
411.6 Overflows in Flush Tanks	405.7.1 Connection of overflows
412.0 Floor Drains and Shower Stalls	905.6 Vent for future fixtures
412.1	412.2 Floor drain trap and strainer
412.2 Location of Floor Drains	
412.2.3	412.4 Public laundries and central washing facilities
412.3 Food Storage Areas	802.1.2 Floor drains in food storage areas
412.4 Floor Slope	

1997 Uniform Plumbing Code	2000 International Plumbing Code
412.5	417.1 Approval 417.4 Shower compartments
412.6	417.5.2 Shower lining
412.7	417.4 Shower compartments
412.8	417.4 Shower compartments 417.5.1 Support 417.5.2 Shower lining 417.5.2.1 PVC sheets 417.5.2.2 Chlorinated polyethylene (CPE) sheets 417.5.2.3 Sheet lead 417.5.2.4 Sheet copper
412.9	417.3 Shower waste outlet
412.11 Location of Valves and Heads	
412.12 Water Supply Riser	417.2 Water supply riser
413.0 Minimum Number of Required Fixtures	
413.1 Fixture Count	403.1 Minimum number of fixtures
413.2 Access to Fixtures	403.5 Location of employee toilet facilities in mercantile and assembly occupancies 403.6 Public facilities 403.6.1 Covered malls
413.2.1	403.4.1 Travel distance
413.3 Separate Facilities	403.1.1 Unisex toilet and bath fixtures
413.4 Fixture Requirements for Special Occupancies	403.1 Minimum number of fixtures
413.5 Facilities in Mercantile and Business Occupancies Serving Customers	403.5 Location of employee toilet facilities in mercantile and assembly occupancies 403.6 Public facilities 403.6.1 Covered malls
413.6 Food Service Establishments	403.4 Location of employee toilet facilities in occupancies other than assembly or mercantile
413.7 Toilet Facilities for Workers	311.1 General
414.0 Fixtures for the Handicapped	404.2.2 Required fixtures 404.2.2.1 Unisex toilet rooms 404.2.2.2 Unisex bathing rooms 404.2.3 Location 404.2.3.1 Prohibited location 404.2.4 Clear floor space 404.2.5 Privacy 404.2.6 Signage

1997 Uniform Plumbing Code	2000 International Plumbing Code
415.0 Whirlpool Bathtubs	
415.1	421.2 Installation
415.2	421.2 Installation
415.3	421.3 Drain
415.4	421.4 Suction fittings
416.0 Installation of Fixture Fittings	
417.0 Bidets	
417.1 Materials	408.1 Approval
417.2 Backflow Protection	408.2 Water connection
419.0 Future Fixtures	704.4 Future fixtures 710.2 Future fixtures
420.0 Shower and Tub/Shower Combination Control Valves	424.4 Shower valves
Table 4-1	Table 403.1 403.3 Number of occupants of each sex
Table 4-1, footnote 3	410.2 Prohibited location
Table 4-1, footnote 5	419.2 Substitution for water closets
Table 4-1, footnote 10	419.3 Surrounding material
Table 4-1, footnote 12	410.1 Approval

Chapter 5: Water Heaters

1997 Uniform Plumbing Code	2000 International Plumbing Code
501.0 General	501.1 Scope
502.0 Definitions	
503.0 Permits	106.1 When required
504.0 Inspection	
504.1 Inspection of Chimneys or Vents	
504.2 Final Water Heater Inspection	
505.0 Gas-Fired Water Heater Approval Requirements	

1997 Uniform Plumbing Code	2000 International Plumbing Code
505.1	501.5 Water heater labeling
505.2	105.5 Material and equipment reuse
506.0 Oil-Burning and Other Water Heaters	
507.0 Combustion Air	
508.0 Clearances	
509.0 Prohibited Locations	502.4 Prohibited location
510.0 Protection from Damage	502.2 Water heaters installed in garages
510.1	
510.2	
510.3	305.9 Protection of components of plumbing system
511.0 Access and Working Space	501.4 Location 502.5 Water heaters installed in attics
512.0 Venting of Water Heaters – General	
513.0 Limitations	
514.0 Vent Connectors	
Table 5-1	
515.0 Location and Support of Venting Systems	
516.0 Length, Pitch, and Clearance	
517.0 Vent Termination	
Table 5-2	
518.0 Area of Venting System	
519.0 Multiple Appliance Venting	
520.0 Existing Venting System	
521.0 Draft Hoods	
522.0 Gas Venting into Existing Masonry Chimneys	
523.0 Chimney Connectors	
524.0 Mechanical Draft Systems	

1997 Uniform Plumbing Code	2000 International Plumbing Code
525.0 Venting Through Ventilating Hoods and Exhaust Systems	

Chapter 6: Water Supply and Distribution

1997 Uniform Plumbing Code	2000 International Plumbing Code
601.0 Running Water Required	602.2 Potable water required
601.1	602.1 General 301.4 Connections to water supply 608.2 Plumbing fixtures
601.2 Identification of a Potable and Nonpotable Water System	608.8 Identification of potable and nonpotable water 608.8.1 Information 608.8.2 Color 608.8.3 Size
Table 6-1	608.8.3 Size Table 608.8.3
602.0 Unlawful Connections	608.17 Protection of individual water supplies
602.1	608.1 General 608.2 Plumbing fixtures 608.3 Devices, appurtenances, appliances and apparatus
602.2	
602.3	608.5 Chemicals and other substances
602.4	608.2 Plumbing fixtures 608.3 Devices, appurtenances, appliances and apparatus 608.5 Chemicals and other substances 608.6.1 Private water supplies
Table 6-2	Table 608.1
603.0 Cross-Connection Control	406.2 Water connection 408.2 Water connection 409.2 Water connection 608.1 General 608.6 Cross-connection control
603.1 Approval of Devices or Assemblies	605.4.1 Dual check-valve-type backflow preventor
603.2 Backflow Prevention Devices, Assemblies, and Methods	
603.2.1 Airgap	608.13.4 Barometric loop 802.2.1 Air gap

1997 Uniform Plumbing Code	2000 International Plumbing Code
603.2.2 Atmospheric Vacuum Breaker (AVB)	608.13.6 Atmospheric-type vacuum breakers
603.2.3 Double Check Valve Backflow Prevention Assembly (DC)	608.13.7 Double check-valve assemblies
603.2.4 Pressure Vacuum Breaker Backflow Prevention Assembly (PVB)	608.13.5 Pressure-type vacuum breakers
603.2.5 Pressure Vacuum Breaker Spill-Proof Type Backflow Prevention Assembly (SVB)	608.13.8 Spill-proof vacuum breakers
603.2.6 Reduced Pressure Principle Backflow Prevention Assembly (RP)	608.13.2 Reduced pressure principle backflow preventers
Table 6-3	608.13.4 Barometric loop
603.3 General Requirements	
603.3.1	605.6 Fittings
603.3.2	312.9 Inspection and testing of backflow prevention assemblies 312.9.1 Inspections 312.9.2 Testing
603.3.3	608.14 Location of backflow preventers 609.6 Clinical, hydrotherapeutic and radiological equipment
603.4 Specific Requirements	
603.4.1 Water Closet and Urinal Flushometer Valves	425.3 Flushometer valves and tanks
603.4.2 Water Closet and Urinal Tanks	425.4 Flush tanks
603.4.3 Water Closet Flushometer Tanks	425.3 Flushometer valves and tanks 608.15.4.2 Hose connections 609.4 Vacuum Breaker Installation
603.4.4 Heat Exchangers	608.16.3 Heat exchangers
603.4.5 Inlets to Tanks, Vats, Sumps, Swimming Pools	608.16.6 Connections subject to back pressure
603.4.6 Protection from Lawn Sprinklers and Irrigation Systems	608.16.5 Connections to lawn irrigation systems
603.4.6.4	608.16.7 Chemical dispensers
603.4.8 Faucets with Pull-Out Spouts	608.15.4.3 Fittings with hose-connected outlets
603.4.9 Water Cooled Compressors, Degreasers	608.3 Devices, appurtenances, appliances and apparatus

1997 Uniform Plumbing Code	2000 International Plumbing Code
603.4.10 Water Inlets to Water Supplied Aspirators	608.3 Devices, appurtenances, appliances and apparatus 609.6 Clinical, hydrotherapeutic and radiological equipment
603.4.11 Potable Water Make Up Connections to Steam or Hot Water Boilers	608.16.2 Connection to boilers
603.4.12 Non-potable Water Piping	608.16.6 Connections subject to back pressure
603.4.13 Potable Water Supply to Carbonators	608.16.1 Beverage dispensers
603.4.14 Water Treatment Units	608.16.7 Chemical dispensers 611.2 Reverse osmosis systems
603.4.15 Backflow Preventers	608.14 Location of backflow preventers 608.15.4 Protection by a vacuum breaker
603.4.16 Deck-Mounted, Equipment-Mounted Atmospheric and Spill-Proof Pressure-Type Vacuum Breakers	608.13.6 Atmospheric-type vacuum breakers 608.13.8 Spill-proof vacuum breakers 605.15.4.1 Deck-mounted and integral vacuum breakers 608.16.4 Connections to automatic fire sprinkler systems and standpipe systems
603.4.17 Faucets with Hose-Attached Sprays	608.15.4.3 Fittings with hose-connected outlets
603.4.18 Protection from Fire Systems	608.16.4 Connections to automatic fire sprinkler systems and standpipe systems
603.4.20 Portable cleaning equipment, dental vacuum pumps	608.16.8 Portable cleaning equipment 608.16.9 Dental pump equipment
604.0 Materials	
604.1	605.4 Water service pipe 605.5 Water distribution pipe
604.2	
604.3	
604.4	
604.5	
604.6	
604.7	105.5 Material and equipment reuse
604.8	601.3 Existing piping used for grounding 605.4 Water service pipe 605.5 Water distribution pipe

1997 Uniform Plumbing Code	2000 International Plumbing Code
604.9	
604.10	605.3 Lead content of water supply pipe and fittings
605.0 Valves	605.7 Valves
605.1	
605.2	606.1 Location of full-open valves
605.3	606.2 Location of shutoff valves 606.3 Access to valves
605.4	606.1 Location of full-open valves
605.5	606.2 Location of shutoff valves
605.6	606.3 Access to valves
605.7	
606.0 Joints and Connections	
606.1 Types of Joints	
606.1.1 Flared Joints	605.15.2 Flared joints
606.1.3 Mechanically Formed Tee Fittings	605.6.1 Mechanically formed tee fittings 605.6.1.1 Full flow assurance
606.2 Use of Joints	
606.2.1 Copper Water Tube	605.15 Copper tubing 605.15.1 Brazed joints 605.15.3 Mechanical joints 605.15.4 Soldered joints
606.2.2 Plastic Fittings	605.9 Prohibited joints and connections 605.22.2 Plastic pipe or tubing to other piping material
607.0 Gravity Supply Tanks	606.5.3 Covers 607.4 Hot water supply to fixtures
608.0 Water Pressure, Pressure Regulators, Pressure Relief Valves, and Vacuum Relief Valves	
608.1 Inadequate Water Pressure	604.7 Inadequate water pressure 606.5 Water pressure booster systems 606.5.1 Water pressure booster systems required

1997 Uniform Plumbing Code	2000 International Plumbing Code
608.2 Excessive Water pressure	604.8 Water-pressure reducing valve or regulator 604.8.2 Repair and removal 607.3.1 Pressure-reducing valve
608.3	607.3 Thermal expansion control 607.3.2 Backflow prevention device or check valve
608.4	
608.5	
608.6	
608.7 Vacuum Relief Valves	504.2 Vacuum relief valves
609.0 Installation, Testing, Unions, and Location	
609.1 Installation	305.6 Freezing
609.2	603.2 Separation of water service and building drain/sewer 608.4 Water service piping
609.3	
609.4 Testing	312.5 Water supply system test 601.4 Tests 606.6 Water supply system test
609.5 Unions	
609.6 Location	
609.7	
609.8 Low Pressure Cutoff Required on Booster Pumps for Water Distribution Systems	606.5.5 Low-pressure cutoff required on booster pumps
Table 6-4	Table 604.3 Appendix Table E-101B
Table 6-5	604.3 Water distribution system design criteria Appendix E
609.9 Disinfection of Potable Water System	601.4 Tests 610.1 General
609.10 Water Hammer	604.9 Water hammer
609.10.1 Air Chambers	604.9 Water hammer
609.10.2 Mechanical Devices	604.9 Water hammer

1997 Uniform Plumbing Code	2000 International Plumbing Code
610.0 Size of Potable Water Piping	604.3 Water distribution system design criteria Appendix E
610.1	604.1 General
610.2	
610.3	
610.4	
610.5	
610.6	
610.7	
610.7(5)	604.6 Variable street pressures
610.8 Size of Meter and Building Supply Pipe Using Table 6-5	603.1 Size of water service pipe
610.9 Size of Branches	604.3 Water distribution system design criteria Appendix E
610.10 Sizing for Flushometer Valves	604.3 Water distribution system design criteria Appendix E
Table 6-6	604.3 Water distribution system design criteria Appendix E
610.11 Sizing Systems For Flushometer Tanks	604.3 Water distribution system design criteria Appendix E
610.12 Sizing Systems with Hot Water Piping	604.3 Water distribution system design criteria Appendix E
610.13	
610.14 Exceptions	102.2 Existing installations 102.4 Additions, alterations or repairs 105.4 Alternative engineered design
611.0 Water Treatment Units	611.1 Design

Chapter 7: Sanitary Drainage

PART 1 – DRAINAGE SYSTEMS	
701.0 Materials	Table 702.1 Above Ground Drainage and Vent Pipe

1997 Uniform Plumbing Code	2000 International Plumbing Code
701.1	702.1 Above-ground sanitary drainage and vent pipe 702.2 Underground building sanitary drainage and vent pipe
701.1.3 The distance below ground for clay pipe shall be at least 12 inches	Table 702.2 Underground Building Drainage and Vent Pipe
701.2	702.4 Fittings
701.3 Lead	402.4 Sheet lead
701.4 Ferrules and Bushings	
701.4.1	705.18 Caulking ferrules
701.4.2	705.19 Soldering bushings
Table 7-1	412.3 Size of floor drains Table 705.18
Table 7-2	Table 705.19
Table 7-3	413.2 Domestic food waste grinder waste outlets 413.3 Commercial food waste grinder waste outlets 416.1 Approval Table 710.1(2) 417.3 Shower waste outlet Table 709.1
702.0 Fixture Unit Equivalents	709.1 Values for fixtures 709.2 Fixtures not listed in Table 709.1 Table 709.2
Table 7-4	709.1 Values for fixtures 709.2 Fixtures not listed in Table 709.1 Table 709.2
703.0 Size of Drainage Piping	Table 710.1(1) Building Drains and Sewers
703.1	710.1 Maximum fixture unit load
703.2	709.3 Values for continuous and semicontinuous flow 710.1 Maximum fixture unit load
703.3	
704.0 Fixture Connections (Drainage)	706.1 Connections and changes in direction
704.1	
704.2	

1997 Uniform Plumbing Code	2000 International Plumbing Code
704.3	409.3.2 Commercial dishwashing machines 802.1.1 Food handling
704.4 Closet Rings (Closet Flanges)	405.4.1 Floor flanges
704.4.5	405.4.2 Securing floor outlet fixtures
Table 7-5 Maximum Unit Loading and Maximum Length of Drainage and Vent Piping NOTE: Three water closets allowed on a 3-inch horizontal branch or building sewer	Table 710.1(1) Building Drains and Sewers Table 710.1(2) Horizontal Fixtures and Stacks 903.5 Vent headers 916.1 Size of stack vents and vent stacks Table 916.1 916.4 Multiple branch vents 916.4.1 Multiple branch vents exceeding 40 feet in developed length
705.0 Joints and Connections	
705.1 Types of Joints	705.1 General
705.1.1 Caulked Joints	705.5.1 Caulked joints
705.1.2 Cement Mortar Joints	705.15 Vitrified clay
705.1.3 Burned Lead Joints	705.13.1 Burned
705.1.4 Asbestos Cement Sewer Pipe Joints	705.3 Asbestos-cement
705.1.6 Molded Rubber Coupling Joints	705.5.3 Mechanical joint coupling
705.1.7 Elastomeric Gasketed and Rubber-Ring Joints	705.5.2 Compression gasket joints 705.5.3 Mechanical joint coupling
705.1.8 Shielded Coupling Joints	705.5.3 Mechanical joint coupling
705.1.9 Hubless Cast Iron Pipe Joints	705.5.3 Mechanical joint coupling
705.2 Use of Joints	
705.2.1 Clay and Sewer Pipe	705.15 Vitrified clay 705.16 Joints between different materials
705.2.2 Cast Iron Pipe	705.5 Cast iron
705.2.3 Screw Pipe to Cast Iron	705.16 Joints between different materials
705.2.4 Lead to Cast Iron, Wrought Iron or Steel	705.16.3 Cast-iron pipe to galvanized steel or brass pipe 705.16.5 Lead pipe to other piping material
705.3 Special Joints	
705.3.1 Slip Joints	705.17 Drainage slip joints

1997 Uniform Plumbing Code	2000 International Plumbing Code
705.3.2 Expansion Joints	308.8 Expansion joint fittings
705.3.3 Ground Joint, Flared or Ferrule Connections	705.18 Caulking ferrules
706.0 Changes in Direction of Drainage Flow	Table 706.3 Fittings for Change of Direction 706.1 Connections and changes in direction 706.3 Installation of fittings
706.1	
706.2	706.3 Installation of Fittings
706.3 Horizontal to horizontal changes must use 45 degree fittings	Table 706.3 Fittings for Change of Direction, footnote a
706.4 The fitting at the base of the stack shall be a 45 degree wye type	Table 706.3 Fittings for Change In Direction
707.0 Cleanouts	708.1 Scope
707.1	708.2 Cleanout plugs
707.2	708.2 Cleanout plugs
707.3	
707.4	708.3.3 Changes of direction 708.3.4 Base of stack
707.4(3)	708.3.1 Horizontal drains within buildings
707.4(4)	708.3.5 Building drain and building sewer junction
707.5 If the total flat turns in the pipe add up to more than 135 degrees, more than one cleanout will be required	708.7 Minimum Size
707.6	708.5 Opening direction
707.7	
707.8	
707.9	708.4 Concealed piping 708.9 Access
707.10	708.8 Clearances
707.11	708.7 Minimum size
707.12	
707.13	708.2 Cleanout plugs

1997 Uniform Plumbing Code	2000 International Plumbing Code
707.14	708.2 Cleanout plugs
Table 7-6	708.7 Minimum size
708.0 Grade of Horizontal Drainage Piping	704.1 Slope of horizontal drainage piping Table 704.1
709.0 Gravity Drainage Required	
710.0 Drainage of Fixtures Located Below the Next Upstream Manhole or Below the Main Sewer Level	1113.1 Pumping system
710.1	715.1 Sewage backflow
710.2	712.1 Building subdrains
710.3	712.3.1 Sump pump
710.4	712.2 Full open valve required 712.3.3 Discharge piping 712.3.5 Ejector connection to the drainage system
710.5	
710.6	715.2 Material 715.3 Seal 715.4 Diameter 715.5 Location
710.7	916.5 Sump vents
710.8	712.3.2 Sump pit
710.9	712.3.4 Maximum effluent level 712.4 Sewage pumps and sewage ejectors
710.10	712.3.2 Sump pit 916.5.1 Sewage pumps and sewage ejectors other than pneumatic
710.11	
710.12	
710.13 Grinder Pump Ejector	712.4.1 Macerating toilet systems
711.0 Suds Relief	704.3 Connections to offsets and bases of stacks 711.1 Horizontal branch connections above or below vertical stack offsets
712.0 Testing	701.6 Tests
712.1 Media	312.1 Required tests

1997 Uniform Plumbing Code	2000 International Plumbing Code
712.2 Water Test	312.2 Drainage and vent water test
712.3 Air Test	312.3 Drainage and vent air test
PART II – BUILDING SEWERS	
713.0 Sewer Required	
713.1	701.2 Sewer required
713.2	701.2 Sewer required
713.3	
713.4	701.2 Sewer required
713.5	
713.6	
714.0 Damage to Public Sewer or Private Sewage Disposal System	701.5 Damage to drainage system or public sewer
715.0 Building Sewer Materials	
715.1	702.3 Building sewer pipe
715.2	
716.0 Markings	
717.0 Size of Building Sewers	710.1 Maximum fixture unit load Table 710.1(2) Horizontal Fixture Branches and Stacks
718.0 Grade, Support, and Protection of Building Sewers	
718.1	
718.2	306.1 Support of piping 306.2 Trenching and bedding 306.2.1 Overexcavation 306.2.3 Soft load-bearing materials
718.3	305.6.1 Sewer depth
719.0 Cleanouts	703.5 Cleanouts on building sewers
719.1	708.3.2 Building sewers 708.3.2 Building sewers
719.2	708.3.5 Building drain and building sewer junction

1997 Uniform Plumbing Code	2000 International Plumbing Code
719.3	708.4 Concealed piping
719.4	708.5 Opening direction
719.5	708.4 Concealed piping
719.6	708.8 Clearances 708.3.2 Building sewers 708.3.6 Manholes
720.0 Sewer and Water Pipes	603.2 Separation of water service and building drain/sewer 608.4 Water service piping 703.1 Building sewer pipe near the water service
721.0 Location	
722.0 Abandoned Sewers and Sewage Disposal Facilities	
723.0 Building Sewer Test	312.6 Gravity sewer test 312.7 Forced sewer test 701.6 Tests
Table 7-7	608.17.1 Well locations Table 608.17.1
Table 7-8 Maximum and Minimum Fixture Unit Loading On Building Sewer Piping	Table 710.1(1) Building Drains and Sewers Table 710.1(2) Horizontal Fixture Branches and Stacks

Chapter 8: Indirect Wastes

1997 Uniform Plumbing Code	2000 International Plumbing Code
801.0 Indirect Wastes	
801.1 Airgap or Airbreak Required	802.2.1 Air gap 409.3.2 Commercial dishwashing machines 802.2.1 Air gap 802.2.2 Air break
801.2 Food and Beverage Handling Establishments	802.1 Where required
801.2.2	802.1.3 Floor drains in food storage areas 802.1.2 Floor drains in food storage areas
801.3 Bar and Fountain Sink Traps	802.1 Where required 802.2 Installation
801.4 Connections from Water Distribution System	802.1.3 Potable clear-water waste
801.5 Sterilizers	802.1.3 Potable clear-water waste

1997 Uniform Plumbing Code	2000 International Plumbing Code
801.6 Drip or Drainage Outlets	802.1 Where required 802.1.5 Nonpotable clear-water waste
801.7 Potable Water Pressure Tanks, Water Treatment Devices, Boilers, and Relief Valves	802.1.3 Potable clear-water waste
802.0 Approvals	
803.0 Indirect Waste Piping	802.2 Installation 901.2.1 Venting required
804.0 Indirect Waste Receptors	802.3 Waste receptors 406.3 Waste connection 802.3 Waste receptors
804.1	412.3 Size of floor drains 802.3.1 Size of receptors 802.4 Standpipes
804.2	
805.0 Pressure Drainage Connections	406.3 Waste connection
806.0 Sterile Equipment	802.1.3 Potable clear-water waste
807.0 Appliances	314.1 Fuel-burning appliances
807.1	406.3 Waste connection 409.3.2 Commercial dishwashing machines
807.2	
807.3	
807.4	406.3 Waste connection 409.3 Waste connection 409.3.1 Domestic dishwashing machines
808.0 Cooling Water	
809.0 Drinking Fountains	
810.0 Steam and Hot Water Drainage Condensers and Sumps	803.1 Waste-water temperature
810.1	701.7 Connections 803.1 Waste-water temperature
810.2	
810.3	
Table 8-1	

1997 Uniform Plumbing Code	2000 International Plumbing Code
810.4 Strainers	802.3 Waste receptors
811.0 Chemical Wastes	702.5 Chemical waste system 803.3 System design
811.1	701.5 Damage to drainage system or public sewer, 803.2 Neutralizing device required for corrosive wastes
811.2	
811.3	
811.4	
811.5	
811.6	901.3 Chemical waste vent system
812.0 Clear Water Wastes	802.1.5 Nonpotable clear-water waste
813.0 Swimming Pools	802.1.4 Swimming pools 802.1.5 Nonpotable clear-water waste
814.0 Refrigeration Wastes	314.2 Evaporators and cooling units 314.2.1 Condensate disposal 314.2.2 Drain pipe material and sizes 802.1.5 Nonpotable clear-water waste
815.0 Air-Conditioning Equipment	314.2 Evaporators and cooling units 314.2.1 Condensate disposal 314.2.2 Drain pipe material and sizes 802.1.5 Nonpotable clear-water waste
815.1 Size	
Table 8-2	
815.2 Point of Discharge	802.2 Installation

Chapter 9: Vents

1997 Uniform Plumbing Code	2000 International Plumbing Code
901.0 Vents Required	901.1 Scope 901.2 Trap seal protection 907.1 Individual vent permitted
902.0 Vents Not Required	901.2.1 Venting required
902.1	
902.2	

1997 Uniform Plumbing Code	2000 International Plumbing Code
903.0 Materials	
903.1	902.1 Vents
903.2 Use of Copper Tubing	902.1 Vents
904.0 Size of Vents	903.1 Stack required
904.1 Size of Vents	903.1 Stack required 916.4 Multiple branch vents
904.2	
904.3	901.6 Engineered systems
905.0 Vent Pipe Grades and Connections	905.2 Grade 906.1 Distance of trap from vent
905.1	905.2 Grade
905.2	905.3 Vent connection to drainage system
905.3	905.4 Vertical rise of vent
905.4	905.1 Connection
905.5	
905.6	908.1 Individual vent as common vent 908.2 Connection at the same level
906.0 Vent Termination	904.1 Roof extension
906.1	903.3 Vent termination 904.1 Roof extension 904.6 Extension through the wall
906.2	904.5 Location of vent terminal 904.6 Extension through the wall
906.3	904.1 Roof extension 904.4 Prohibited use
906.4	904.1 Roof extension
906.5	904.3 Flashings
906.6 Lead	902.3 Sheet lead
906.7 Frost or Snow Closure	904.2 Frost closure
907.0 Vent Stacks and Relief Vents	

1997 Uniform Plumbing Code	2000 International Plumbing Code
907.1	903.2 Vent stack required 903.4 Vent connection at base 914.1 Where required 914.2 Size and connection
907.2	
908.0 Vertical Wet Venting	908.1 Individual vent as common vent
908.1	908.3 Connection at different levels
908.2	908.3 Connection at different levels
908.3	
909.0 Special Venting for Island Fixtures	913.1 Limitation 913.2 Vent connection 913.3 Vent installation below the fixture flood level rim
910.0 Combination Waste and Vent Systems	912.1 Type of fixtures
910.1	
910.2	
910.3	912.2 Installation 912.2.2 Connection 912.2.3 Vent size
910.4	912.3 Size
910.5	912.2 Installation
910.6	
910.7	912.1 Type of fixtures

Chapter 10: Traps and Interceptors

1001.0 Traps Required	1002.1 Fixture traps
1001.1	412.2 Floor drain trap and strainer
1001.2	
1001.3	
1001.4	
1002.0 Traps Protected by Vent Pipes	

1997 Uniform Plumbing Code	2000 International Plumbing Code
1002.1	901.2 Trap seal protection
1002.2	906.1 Distance of trap from vent 906.3 Crown vent
1002.3	
1002.4	
Table 10-1 Horizontal Distance of Trap Arms	Table 906.1 Maximum Distance of Trap From Vent
1003.0 Traps – Described	
1003.1	1002.2 Design of traps 1004.1 General
1003.2	
1003.3	1002.5 Size of fixture traps
1004.0 Traps – Prohibited	1002.2 Design of traps 1002.3 Prohibited traps 1002.6 Building traps
1005.0 Trap Seals	1002.4 Trap seals 1002.7 Trap setting and protection
1006.0 Floor Drain Traps	
1007.0 Trap Seal Protection	1002.4 Trap seals
1008.0 Building Traps	1003.7 Bottling establishments
1009.0 Industrial Interceptors (Clarifiers) and Separators	
1009.1 When Required	1003.1 Where required 1003.6 Laundries
1009.2 Approval	1003.2 Approval
1009.3 Design	1003.5 Sand interceptors in commercial establishments
1009.4 Relief Vent	1003.9 Venting of interceptors and separators
1009.5 Location	1003.10 Access and maintenance of interceptors and separators
1009.6 Maintenance of Interceptors	1003.10 Access and maintenance of interceptors and separators
1009.7 Discharge	1003.4 Oil separators required

1997 Uniform Plumbing Code	2000 International Plumbing Code
1010.0 Slaughter Houses, Packing Establishments, etc.	1003.8 Slaughterhouses
1011.0 Minimum Requirements for Auto Wash Racks	1003.4 Oil separators required
1012.0 Laundries	1003.6 Laundries
1013.0 Bottling Establishments	1003.7 Bottling establishments
1014.0 Grease Traps and Grease Interceptors	1003.3 Grease traps and grease interceptors
1014.1	1003.1 Where required 1003.3.1 Grease traps and grease interceptors required 1003.3.3 Grease trap and grease interceptor not required
1014.2	1003.3.4.1 Grease trap capacity
1014.3	1003.3.4.2 Rate of flow controls
1014.4	1003.3.4.1 Grease trap capacity
1014.5	
1014.6	
1014.7	
1014.8	
1014.9	
1014.10 Grease Interceptors For Commercial Kitchens	
1015.0 Food Waste Disposal and Dishwasher Prohibited	1002.1 Fixture Trap 1003.3.2 Food waste grinders
Table 10-2 Grease Traps	Table 1003.3.4.1 Capacity of Grease Traps
1016.0 Sand Interceptors	
1016.1 Where Required	1003.1 Where required
1016.2 Construction and Size	
1016.3 Separate Use	1003.5 Sand interceptors in commercial establishments
1016.4 Alternate Design	
1017.0 Oil and Flammable Liquids Interceptors	

1997 Uniform Plumbing Code	2000 International Plumbing Code
1017.1 Interceptors Required	1003.4 Oil separators required
1017.2 Design of Interceptors	1003.4.2.1 General design requirements 1003.4.2.2 Garages and service stations
1017.3 Combination Oil and Sand Interceptor	

Chapter 11: Storm Drainage

1997 Uniform Plumbing Code	2000 International Plumbing Code
1101.0 General	
1101.1 Where Required	1101.2 Where required
1101.2 Storm Water Drainage to Sanitary Sewer Prohibited	1101.3 Prohibited drainage
1101.3 Material Uses	1102.1 General 1102.2 Inside storm drainage conductors
1101.4 Expansion Joints Required	305.3 Stress and strain 308.8 Expansion joint fittings
1101.5 Subsoil Drains	1111.1 Subsoil drains
1101.5.3	1113.1.1 Pump capacity and head 1113.1.2 Construction
1101.6 Building Subdrains	1112.1 Building subdrains
1101.7 Areaway Drains	
1101.8 Window Areaway Drains	
1101.9 Filling Station and Motor Vehicle Washing Establishments	1101.2 Where required
1101.10 Paved Areas	1101.2 Where required
1101.11 Roof Drainage	**1106 SIZE OF CONDUCTORS, LEADERS AND STORM DRAINS**
1101.11.1 Primary Roof Drainage	1106.1 General
1101.11.2 Secondary Roof Drainage	1107.1 Secondary drainage required
1101.11.2.1	1107.3 Sizing of secondary drains
1101.11.2.2	1107.2 Separate systems required
1101.11.3 Equivalent Systems	**1110 CONTROLLED FLOW ROOF DRAIN SYSTEMS**
1101.12 Cleanouts	1101.8 Cleanouts required

1997 Uniform Plumbing Code	2000 International Plumbing Code
1102.0 Materials	1102.1 General
1102.1 Conductors	1102.2 Inside storm drainage conductors
1102.2 Leaders	1102.1 General
1102.3 Underground Building Storm Drains	1102.2 Inside storm drainage conductors
1102.4 Building Storm Sewers	1102.4 Building storm sewer pipe
1102.5 Subsoil Drains	1102.5 Subsoil drain pipe 1111.1 Subsoil drains
1103.0 Traps on Storm Drains and Leaders	
1103.1 Where Required	1103.1 Main trap
1103.2 Where Not Required	1101.2 Where required 1103.1 Main trap
1103.3 Trap Size	1103.3 Size
1103.4 Method of Installation of Combined Sewer	1103.4 Cleanout
1104.0 Leaders, Conductors, and Connections	
1104.1 Improper Use	1104.1 Prohibited use
1104.2 Protection of Leaders	305.9 Protection of components of plumbing system
1104.3 Combining Storm with Sanitary Drainage	1104.2 Combining storm with sanitary drainage
1105.0 Roof Drains	
1105.1 Material	1102.6 Roof drains
1105.2 Dome or Strainer for General Use	1105.1 Strainers
Table 11-1	Figure 1106.1 Table 1106.2
Table 11-2	Figure 1106.1 Table 1106.3
Table 11-3	Figure 1106.1 Table 1106.6
1105.3 Strainers for Flat Decks	1105.2 Flat decks
1105.4 Roof Drain Flashings	1105.3 Roof drain flashings
1106.0 Size of Leaders, Conductors, and Storm Drains	1106.1 General

1997 Uniform Plumbing Code	2000 International Plumbing Code
1106.1 Vertical Conductors and Leaders	1106.2 Vertical conductors and leaders
1106.2 Size of Building Storm Drains and Sewers	1106.3 Building storm drains and sewers
1106.3 Size of Roof Gutters	1106.6 Size of roof gutters
1106.4 Side Walls Draining onto a Roof	1106.4 Vertical walls
1107.0 Values for Continuous Flow	**1109 VALUES FOR CONTINUOUS FLOW**
1108.0 Controlled-Flow Roof Drainage	**1110 CONTROLLED FLOW ROOF DRAIN SYSTEMS**
1108.1 Application	
1108.1(3)	1110.4 Minimum number of roof drains
Table 11-4	
Table 11-5	
1108.2 Setback Roofs	**1110 CONTROLLED FLOW ROOF DRAIN SYSTEMS**
1109.0 Testing	312.8 Storm drainage system test 1101.4 Tests
1109.1 Testing Required	312.8 Storm drainage system test 1101.4 Tests
1109.2 Methods of Testing Storm Drainage Systems	312.8 Storm drainage system test 1101.4 Tests
1102.2.1 Water Test	312.8 Storm drainage system test 1101.4 Tests
1109.2.2 Air Test	312.8 Storm drainage system test 1101.4 Tests
1109.2 Exceptions	

Chapter 12: Fuel Piping

1201.0 General	See 2000 *International Fuel Gas Code* for all requirements for fuel gas piping and associated provisions such as; combustion air, venting, water heaters, etc.
1202.0 Definitions	
1203.0 Workmanship	
1204.0 Inspection	

1997 Uniform Plumbing Code	2000 International Plumbing Code
1205.0 Certification of Inspection	
1206.0 Authority to Render Gas Service	
1207.0 Authority to Disconnect	
1208.0 Temporary Use of Gas	
1209.0 Gas Meter Locations	
1210.0 Material for Gas Piping	
1211.0 Installation of Gas Piping	
1212.0 Appliance Connectors	
1213.0 Liquified Petroleum Gas Facilities and Piping	
1214.0 Leaks	
1215.0 Interconnections of Gas Piping Systems	
1216.0 Required Gas Supply	
1217.0 Required Gas Piping Size	
1218.0 Medium Pressure Gas Piping Systems	

Chapter 13: Medical Gas Systems

Chapter 13 – Medical Gas Systems	**713 HEALTH CARE PLUMBING** 1202.1 Nonflammable medical gases (See Comparison Discussion)

Chapter 14: Mandatory Referenced Standards

Table 14-1 Standards for Materials, Equipment, Joints and Connections	**CHAPTER 13 – REFERENCED STANDARDS**
Table 14-2 Fixture and Fitting Standards	**CHAPTER 13 – REFERENCED STANDARDS**

APPENDICES

APPENDIX A – RECOMMENDED RULES FOR SIZING THE WATER SUPPLY SYSTEM	CHAPTER 6 – WATER SUPPLY AND DISTRIBUTION APPENDIX E SIZING OF WATER PIPING SYSTEM

1997 Uniform Plumbing Code	2000 International Plumbing Code
APPENDIX B – EXPLANATORY NOTES ON COMBINATION WASTE AND VENT SYSTEMS	912 COMBINATION DRAIN AND VENT SYSTEM
APPENDIX C – ADDITIONAL REFERENCED STANDARDS	CHAPTER 13 – REFERENCED STANDARDS
APPENDIX D – SIZING STORMWATER DRAINAGE SYSTEMS	APPENDIX B – RATES OF RAINFALL FOR VARIOUS CITIES
Table D-1	APPENDIX B – RATES OF RAINFALL FOR VARIOUS CITIES
Table D-2	
APPENDIX E – MANUFACTURED/MOBILE HOME PARKS AND RECREATIONAL VEHICLE PARKS	
APPENDIX F – (RESERVED)	
APPENDIX G – GRAYWATER SYSTEMS FOR SINGLE FAMILY DWELLINGS	APPENDIX C – GRAY WATER RECYCLING SYSTEMS
APPENDIX H – RECOMMENDED PROCEDURES FOR DESIGN, CONSTRUCTION AND INSTALLATION OF COMMERCIAL KITCHEN GREASE INTERCEPTORS	1003 INTERCEPTORS AND SEPARATORS
APPENDIX I – INSTALLATION STANDARDS	
IS 1-91 NON-METALLIC BUILDING SEWERS	CHAPTER 13 – REFERENCED STANDARDS
IS 2-90 TILE-LINED ROMAN BATHTUBS	
IS 3-93 COPPER PLUMBING TUBE, PIPE AND FITTINGS	CHAPTER 13 – REFERENCED STANDARDS
IS 4-96 TILE-LINED SHOWER RECEPTORS (AND REPLACEMENTS)	
IS 5-92 ABS BUILDING DRAIN, WASTE, AND VENT PIPE AND FITTINGS	CHAPTER 13 – REFERENCED STANDARDS
IS 6-95 HUBLESS CAST IRON SANITARY AND RAINWATER SYSTEMS	CHAPTER 13 – REFERENCED STANDARDS
IS 7-90 POLYETHYLNE (PE) COLD WATER BUILDING SUPPLY	CHAPTER 13 – REFERENCED STANDARDS
IS 8-95 PVC COLD WATER BUILDING SUPPLY AND YARD PIPING	CHAPTER 13 – REFERENCED STANDARDS
IS 9-95 PVC BUILDING DRAIN, WASTE AND VENT PIPE AND FITTINGS	CHAPTER 13 – REFERENCED STANDARDS
IS 10-93 DISCONTINUED	
IS 11-87 ABS SEWER PIPE AND FITTINGS	CHAPTER 13 – REFERENCED STANDARDS

1997 Uniform Plumbing Code	2000 International Plumbing Code
IS 12-93 POLYETHYLENE (PE) FOR GAS YARD PIPING	
IS 13-91 PROTECTIVELY COATED PIPE	
IS 15-82 ASBESTOS CEMENT PRESSURE PIPE FOR WATER SERVICE AND YARD PIPING	CHAPTER 13 – REFERENCED STANDARDS
IS 16-84 LOW PRESSURE AIR TEST FOR BUILDING SEWERS	
IS 18-85 EXTRA STRENGTH VITRIFIED CLAY PIPE IN BUILDING DRAINS	CHAPTER 13 – REFERENCED STANDARDS
IS 20-96 CPVC SOLVENT CEMENTED HOT AND COLD WATER DISTRIBUTION SYSTEMS	CHAPTER 13 – REFERENCED STANDARDS
IS 21-89 WELDED COPPER AND COPPER ALLOY WATER TUBE	CHAPTER 13 – REFERENCED STANDARDS
APPENDIX J – RECLAIMED WATER SYSTEMS FOR NON-RESIDENTIAL BUILDINGS	
APPENDIX K – PRIVATE SEWAGE DISPOSAL SYSTEMS	2000 INTERNATIONAL PRIVATE SEWAGE DISPOSAL CODE
APPENDIX L – ALTERNATE PLUMBING SYSTEMS	105.4 Alternative engineered design **709 FIXTURE UNITS**

APPENDIX B

The following article is reprinted with permission from the May-June 1997 edition of *Building Standards* magazine published by ICBO. The costs and assumptions may have changed since 1997. Note that the 2000 UPC has not accepted air-admittance valves and the cost differences may be more today.

Which is More Cost Effective, IPC or UPC

by Julius Ballanco, P.E.
President
JBEngineering and Code Consulting, P.C.
Munster, Indiana

Julius Ballanco, P.E., is president of JB Engineering and Code Consulting, P.C., in Munster, Indiana. The firm specializes in code and standard consulting for life safety, fire-protection, plumbing and mechanical engineering. Before establishing JB Engineering, Mr. Ballanco was head of Plumbing and Mechanical Engineering for Building Officials and Code Administrators (BOCA) International, Inc. A well-known lecturer and instructor, he is the author of BOCA® National Plumbing Code Commentary *and* Plumbing of Residential Fire Sprinklers *and co-author of* Illustrated National Plumbing Code Design Manual. *A graduate of Stevens Institute of Technology in Hoboken, New Jersey, he is a registered professional engineer and a licensed master plumber, and serves on numerous national standard committees.*

Introduction

The expression "cost is no object" is often heard regarding construction codes. This expression is one of the biggest falsehoods in code enforcement. Cost, in fact, is *always* an object. In a market-driven economy, a code cannot make declarations of greatness while raising the price of construction to unreasonable levels.

Every code must recognize cost impact while maintaining the necessary protection of the public's health, safety and welfare. Codes are constantly accused of being controlled by select individuals that allow only what they consider to be appropriate. An effective code must be open to all innovative ideas and methods, provided it achieves a minimum level of protection.

Installation Cost Comparison

In an evaluation of two model plumbing codes, the *International Plumbing Code*® (IPC) and the *Uniform Plumbing Code*® (UPC), the difference in construction costs and available options was found to be significant. While both codes regulate plumbing system design and installation, the approach that each code takes is completely different. The *International Plumbing Code* allows a variety of options, ranging from acceptable materials to system designs, while the *Uniform Plumbing Code* is more restrictive in material use and system design. A more restrictive plumbing code normally translates into a higher cost of construction, but the only viable way to make that determination is by performing a cost comparison of the installation of a plumbing system in two identical buildings. Such an analysis was performed on a single-family dwelling, with one plumbing system designed to the IPC and the other designed to the UPC.

To ensure impartiality, the layout of the single-family dwelling was randomly chosen from a "build-your-own-home" magazine. The dwelling selected was a two-and-a-half bath, four bedroom house (Figure 1). The plumbing facilities included three water closets, four lavatories, two bathtubs, a kitchen sink and an automatic clothes washer standpipe. The gas utilization appliances included a kitchen range, furnace, water heater, dryer and gas log lighter. The design employed the lowest-cost plumbing system that would be permitted for each building.

It is important to note that while the material and methods were the lowest cost for this particular building, other materials and designs would have been acceptable. *Endorsement of any particular design or material was not intended when performing this comparison.* Additionally, as technology advances and material and labor costs change, the lowest-cost plumbing installation also changes.

The method of estimating the cost of construction follows the National Labor Estimator and the National Association of Plumbing, Heating and Cooling Contractors Labor Calculator. The material prices were obtained from a local plumbing supply house in April 1997. The prices are the wholesale prices charged to plumbing contractors. Contractors who purchase large volumes may receive larger discounts than the prices reflected in the pricing tables, and in different parts of the country, supply houses may have different prices. The time of installation, or increments of labor, were taken from the referenced documents.

The cost comparison was performed only for the piping systems, which included the drainage, waste and venting system (DWV); the water distribution system; and the gas piping system. The cost of installing the fixtures, traps and appliances would be the same under both codes because the layout is identical. Only the piping systems would have an impact on the cost differential of the plumbing system.

DWV System

Figure 2 shows the layout of the DWV system that conforms to the IPC. The cost-cutting measures used in this design are a combination of wet venting the bathroom groups and terminating the vents to air admittance valves.

Section 909 of the IPC permits two bathroom groups to be combined for venting purposes. While the drainage pipe may increase in size for this system, the amount of piping is greatly reduced. The other reduction in piping results from terminating the vents to an air admittance valve, permitted in IPC Section 917. When air admittance valves are used, one vent must extend to the outdoors. The vent for the automatic clothes washer standpipe was selected because the fixture is located in the one-story area of the dwelling unit.

Neither the combined wet venting nor air admittance valves are permitted by the UPC. Additionally, the IPC permits the drain for the kitchen sink to be $1^1/_2$ inches (38 mm) in diameter while the UPC requires a 2-inch (51 mm) drain.

Figure 3 depicts the DWV system designed in conformance to the UPC. What is immediately obvious is the additional vent piping that is required for the DWV system; the only combined venting is located on the second floor. The lavatories are

TABLE 1—DWV MATERIAL AND LABOR, IPC DESIGN

Material	Quantity	Price (each, $ per ft.)	Labor (hours)	Total Labor (hours)	Total Cost of Material ($)
$1^1/_2$" PVC Pipe	67'	0.26	0.05	3.35	17.42
$1^1/_2$" PVC Quarter Bend	6	0.26	0.28	1.68	1.58
$1^1/_2$" PVC Sanitary Tee	4	0.47	0.42	1.68	1.89
2" PVC Pipe	30'	0.35	0.06	1.8	10.50
2" PVC Quarter Bend	2	0.33	0.4	0.8	0.67
2" PVC Sanitary Tee	1	0.67	0.6	0.6	0.67
2" x $1^1/_2$" PVC Double Sanitary Tee	1	1.69	0.6	0.6	1.69
2" x $1^1/_2$" Increaser	1	0.22	0.4	0.4	0.22
3" PVC Pipe	50'	0.70	0.09	4.5	35.00
3" PVC Quarter Bend	5	1.04	0.6	3	5.18
3" PVC Tee-Wye	4	3.21	0.9	3.6	12.84
3" Floor Flange	3	1.74	0.7	2.1	5.22
3" x $1^1/_2$" PVC Tee-Wye	4	3.74	0.9	3.6	14.96
3" x 2" PVC Tee-Wye	1	2.34	0.85	0.85	2.34
3" x 2" Increaser	2	0.73	0.6	1.2	1.45
Roof Flashing	1	3.94	0.75	0.75	3.94
$1^1/_2$" Air Admittance Valve	4	15.45	0.17	0.68	61.80
Total Labor Time and Material Cost for DWV System				**31.19**	**$177.37**

TABLE 2—DWV MATERIAL AND LABOR, UPC DESIGN

Material	Quantity	Price (each, $ per ft.)	Labor (hours)	Total Labor (hours)	Total Cost of Materials ($)
1 1/2" PVC Pipe	80'	0.26	0.05	4	20.80
1 1/2" PVC Quarter Bend	11	0.26	0.28	3.08	2.90
1 1/2" PVC Sanitary Tee	4	0.47	0.42	1.68	1.89
1 1/2" PVC Double Sanitary Tee	1	0.87	0.42	0.42	0.87
2" PVC Pipe	75'	0.35	0.06	4.5	26.25
2" PVC Quarter Bend	4	0.33	0.4	1.6	1.33
2" PVC Sanitary Tee	2	0.67	0.6	1.2	1.35
2" x 1 1/2" PVC Sanitary Tee	1	0.67	0.6	0.6	0.67
2" PVC Tee-Wye	3	1.95	0.6	1.8	5.84
2" x 1 1/2" Increaser	1	0.22	0.4	0.4	0.22
3" PVC Pipe	73'	0.70	0.09	6.57	51.10
3" PVC Quarter Bend	9	1.04	0.6	5.4	9.32
3" PVC Sanitary Tee	2	1.79	0.9	1.8	3.57
3" x 1 1/2" PVC Sanitary Tee	2	1.37	0.9	1.8	2.74
3" PVC Tee-Wye	4	3.21	0.9	3.6	12.84
3" Floor Flange	3	1.74	0.7	2.1	5.22
3" x 1 1/2" PVC Tee-Wye	1	3.74	0.9	0.9	3.74
3" x 2" PVC Tee-Wye	5	2.34	0.9	4.5	11.69
3" x 2" Increaser	2	0.73	0.6	1.2	1.45
Roof Flashing	1	3.94	0.75	0.75	3.94
Total Labor Time and Material Cost for DWV System				**47.9**	**$167.73**

TABLE 3—WATER DISTRIBUTION SYSTEM, IPC DESIGN

Material	Quantity	Price (each, $ per ft.)	Labor (hours)	Total Labor (hours)	Total Cost of Materials ($)
3/8" PEX pipe	550'	0.21	0.01	5.5	115.50
1/2" PEX pipe	30'	0.41	0.02	0.6	12.30
3/8" PEX to Faucet Connector Fitting	19	0.57	0.16	3.04	10.87
Manibloc-Manifold	1	82.47	1.84	1.84	82.47
Total Labor Time and Material Cost for Water Distribution System				**10.98**	**$221.14**

common vented, as permitted in UPC Section 905.6. The water closet, lavatory and bathtub are vented by a vertical wet vent, as permitted in UPC Section 908.0. The *Uniform Plumbing Code* does not permit wet venting on the horizontal plane.

A bill of materials and time increments for installation were prepared for both designs. Table 1 indicates a material cost of $177.37, requiring 31.19 hours of labor, to install the DWV system designed to the IPC. Table 2 indicates a material cost of $167.73, requiring 47.9 hours of labor, for the UPC DWV system design. While the cost of materials is lower for the UPC design, the labor required is almost 54 percent greater than the labor required for the system designed to the IPC.

Water Distribution System

The cost of installing a water distribution system varies based on the type of material permitted by the codes. Section 605.5 of the IPC permits the use of cross-linked polyethylene (PEX) plastic tubing. One of the savings of installing PEX tubing is the opportunity to install a manifold piping system, which has a central manifold with separate hot and cold water lines to each fixture. While this system uses more pipe, the amount of time needed to install it is greatly reduced.

Table 14-1 of the UPC permits the use of Chlorinated Polyvinyl Chloride (CPVC) plastic pipe for water distribution systems. A system piped with CPVC would not use a manifold system; the system would be piped throughout the dwelling unit with take-offs for each fixture. This system would use less piping, but the labor required for installation would increase.

Table 3 indicates the cost of installing a manifold system in accordance with the IPC. The material costs are $221.14, while the labor required for installation is 10.98 hours. Table 4 shows the costs for the water distribution system designed to the UPC as $115.70 for material, requiring 18.92 hours of labor.

For the water distribution system, the material cost using the manifold system permitted by the IPC is almost double, but the labor required to install the system is cut almost in half. Similar to the DWV system, the savings come from lower labor costs, not material costs.

Gas Piping System

In most single-family dwellings, the gas piping is installed as a part of the plumbing. The *International Mechanical Code®* (IMC) regulates gas piping installations; these requirements are duplicated in IPC Appendix G. The *Uniform Plumbing Code* regulates gas piping installation in Chapter 12.

One of the significant differences between the two codes is the material acceptable for gas piping installation. The IPC (IMC) permits the use of corrugated stainless steel tubing (CSST), which is not recognized in the UPC; therefore, a gas piping system installed in accordance with the UPC would have to use black steel pipe.

A benefit of using CSST is that a manifold system can be installed. While this system would use more piping material, again, the savings in labor would reduce the cost of the installation.

Table 5 indicates that the material cost was $225.33, with 9.16 hours in labor, to install the CSST system designed in accordance with the IPC (IMC). The material cost and labor for the gas piping system according to the UPC, shown in Table 6, was $96.94 and 19.83 hours, respectively. When using the IPC, the material costs for CSST are more than double, but the labor required to install the system is less than half.

TABLE 4—WATER DISTRIBUTION SYSTEM, UPC DESIGN

Material	Quantity	Price (each, $ per ft.)	Labor (hours)	Total Labor (hours)	Total Cost of Materials ($)
1/2" CPVC Pipe	55'	0.23	0.03	1.65	12.82
1/2" CPVC Elbows	9	0.10	0.16	1.44	0.86
1/2" CPVC Tees	5	0.14	0.24	1.2	0.68
3/4" CPVC Pipe	165'	0.42	0.03	4.95	69.63
3/4" CPVC Elbows	17	0.22	0.16	2.72	3.66
3/4" CPVC Tees	12	0.31	0.24	2.88	3.66
Angle Stop and Supply	17	1.44	0.24	4.08	24.41
Total Labor Time and Material Cost for Water Distribution System				18.92	$115.70

TABLE 5—GAS PIPING SYSTEM, IPC (IMC) DESIGN

Material	Quantity	Price (each, $ per ft.)	Labor (hours)	Total Labor (hours)	Total Cost of Material ($)
1" Steel Pipe	10'	0.98	0.06	0.6	9.80
1" Elbow	1	1.02	0.56	0.56	1.02
3/4" CSST	15'	1.74	0.05	0.75	26.10
1/2" CSST	105'	1.34	0.05	5.25	140.70
CSST Appliance Connector	5	7.10	0.3	1.5	35.50
5-Port Manifold for CSST	1	12.21	0.5	0.5	12.21
Total Labor Time and Material Cost for Gas Piping System				9.16	$225.33

TABLE 6—GAS PIPING SYSTEM, UPC DESIGN

Material	Quantity	Price (each, $ per ft.)	Labor (hours)	Total Labor (hours)	Total Cost of Material($)
$1/2''$ Steel Pipe	82'	0.59	0.05	4.1	48.38
$1/2''$ Elbow	11	0.48	0.46	5.06	5.28
$1/2''$ Coupling	3	0.90	0.46	1.38	2.70
$1/2''$ Union	4	2.13	0.6	2.4	8.52
$3/4''$ Steel Pipe	15'	0.70	0.05	0.75	10.50
$3/4''$ Elbow	2	0.58	0.5	1	1.16
$3/4''$ x $1/2''$ Tee	2	1.38	0.73	1.46	2.76
$3/4''$ x $3/4''$ Reducer	1	0.87	0.5	0.5	0.87
$3/4''$ Union	1	2.45	0.6	0.6	2.45
1" Steel Pipe	10'	0.98	0.06	0.6	9.80
1" Elbow	1	1.02	0.56	0.56	1.02
1" x $3/4''$ Tee	1	1.75	0.73	0.73	1.75
1" x $1/2''$ Tee	1	1.75	0.69	0.69	1.75
Total Labor Time and Material Cost for Gas Piping System				**19.83**	**$96.94**

TABLE 7—COMPARISON OF LABOR AND MATERIAL COSTS

Type of System	International Plumbing Code		Uniform Plumbing Code	
	Labor (hours)	Material Cost ($)	Labor (hours)	Material Cost ($)
DWV System	31.19	177.37	47.9	167.73
Water Distribution	10.98	221.14	18.92	115.70
Gas Piping	9.16	225.33	19.83	96.94
TOTAL	**51.33**	**623.85**	**86.65**	**$380.38**

Total Cost of Piping Installation

The total costs of materials and labor required to install the piping systems are listed in Table 7. The obvious difference is in the labor required to install the lowest-cost system designed in accordance with the *Uniform Plumbing Code*. Using the UPC would require 35 more hours' labor to install the plumbing piping systems than using the *International Plumbing Code*. Note, however, that although the IPC provides labor savings, the material cost increases by $243.

TABLE 8—COST COMPARISON BASED ON DIFFERENT LABOR PRICE

Labor Rate	International Plumbing Code	Uniform Plumbing Code	UPC Increase (percent)
$20 Per Hour	$1,650.45	$2,113.38	28.05
$35 Per Hour	$2,420.40	$3,413.13	41.02
$50 Per Hour	$3,190.35	$4,712.88	47.72
$65 Per Hour	$3,960.30	$6,012.63	51.82

A thorough comparison of the total cost of the piping system must include the cost of labor. One of the problems in comparing costs is that labor prices fluctuate from region to region. Most plumbing professionals currently would agree that the national labor price average is $35 per hour. Table 8 provides a comparison of costs for four labor prices, ranging from $20 per hour to $65 per hour. In all cases, it is less expensive to install a plumbing piping system in accordance with the IPC than to install the same system in accordance with the UPC. At the national average of $35 per hour, using the *Uniform Plumbing Code* results in an increase of 41 percent in the price of the system.

Conclusion

When a cost comparison of one single-family dwelling is performed, it naturally raises questions of whether the systems were rigged to favor one code over the other. In subsequent evaluations of numerous other building layouts, in all cases, the plumbing system designed in accordance with the *International Plumbing Code* resulted in lower costs. The larger the building, the greater the difference in the cost of the system.

Another factor that cannot be ignored is that not all plumbing systems are designed at the lowest cost to the building owner. That does not, however, diminish the fact that the IPC provides the option for a lower-cost system. It is important to note that both codes require systems that perform as required.

If it could be proven that a system permitted by one code would fail to perform, then there would a valid reason for concern. There is no merit, however, to any of the arguments regarding the validity of the plumbing systems that result in lower costs (as permitted in the IPC). Careful research of all the technical documentation and field investigations of the various systems permitted by both codes has concluded that the level of protection of public health, safety and welfare is maintained. Neither code will jeopardize the public nor compromise the level of protection.

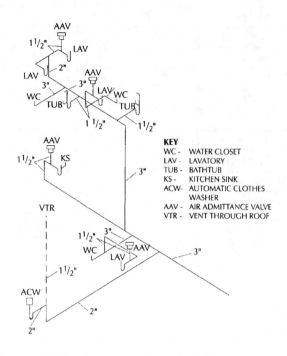

Figure 2—DWV system conforming to the IPC.

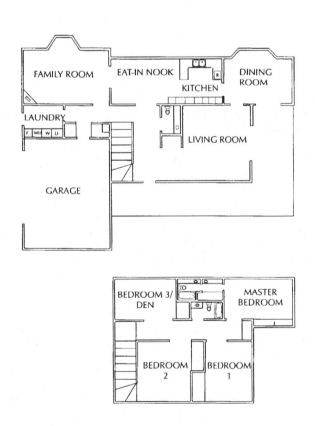

Figure 1—Layout of single-family dwelling in cost comparison.

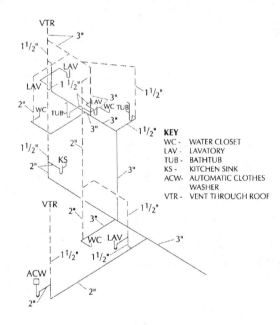

Figure 3—DWV system conforming to the UPC.

INDEX

A

ABS	13
Absorption trench	13
Accepted engineering practice	13
Access cover	14
Access door	14
Accessibility	5
Accessible	14
Acid waste piping	14
Acrylonitrile-butadiene-styrene (ABS)	14
Adapter fitting	14
Administration	4
Administrative authority	14
Administrative details	9
Applicability	10
Building or plumbing departments	9
Permits	9
Purpose of the *International Plumbing Code*	9
Aerator	14
Air break	14
Air gap	6
Air gap (drainage system)	14
Air gap (water-distribution system)	14
Air test	14
Air-admittance valve	14
Alternative engineered design	14
Anchors	14
Angle stop	14
Angle valve	14
Anneal	14
Anode	14
Antisiphon	14
Approvals	32
Approved	14
Area drain	14
Asbestos	14
Asphaltic solution	14
Aspirator	14

B

Backfill	14
Backflow	14
Backpressure	14
Backsiphonage	14
Backwater valve	14
Drainage	15
Water Supply System	15
Backflow connection	15
Backflow preventer	15
Backflow protection	6
Background	1
Back-to-back water closets	6
Bacteria (aerobic)	15

Baffle	15
Ball cock	15
Basin	15
Bathroom group	15
Battery fixtures	15
Beam clamp	15
Bedpan washer	15
Bell and spigot joint	15
Bidet	15
Black iron pipe	15
Boiler drain	16
Bonnet	16
Branch	16
Branch interval	16
Branch vent	16
Brazing	16
Btu	16
Building	16
Building drain	16
Building sewer	16
Building subdrain	16
Building trap	16
Bushing	16
Bypass	16

C

Cap	16
Capillary action	16
Carrier	16
Cast iron	16
Catch basin	16
Cesspool	16
Check valve	16
Circuit vent	16
Cistern	16
Clarifier	16
Cleanout	7, 16, 47
Closed system	17
Closet flange	17
Closet gasket	17
Code	17
Code change process	3
Code official	17, 32
Combination fixture	17
Combination waste and vent system	17
Combined	17
Common vent	17
Condensate drain	17
Conductor	17
Confined space	17
Construction documents	30
Contamination	17
Continuous waste	17
Corporation cock	17

141

INDEX

Coupling 17
Critical Level (C-L) 17
Cross connection 17
Crown weir 17

D

Dead end 17
Definitions 13-27
Depth of water seal 17
Developed length 17
Development of the code 2
Dewpoint 17
Dip of trap 17
Discharge pipe 18
Dishwashers 74
Distribution system 18
Downspout 18
Drain 18
Drain sizing 46
Drainage fittings 18
Drainage fixture unit (dfu) 18
Drainage system 18
    Building gravity 18
    Sanitary 18
    Storm 18
Drainline connections 43
Dry well 18
Ductile iron 18
DWV—installation 39

E

Effective opening 18
Effluent 18
Ejector 18
Electrolysis 18
Emergency floor drain 18
Enamel 18
Enforcement 4
Escutcheon 18
Essentially nontoxic transfer fluids 18
Essentially toxic transfer fluids 18
Existing installations 18
Expansion joint 18

F

Faucet 19
Federal guidelines 4
Ferrule 19
Filter system 19
Fitting 19
Fixture 19, 72
Fixture branch 19
Fixture drain 19
Fixture fitting 19
    Supply fitting 19
    Waste fitting 19

Fixture requirements 4
Fixture supply 19
Fixture units 6
Flaring system 19
Flash point 19
Flashing 19
Flood zones 19
    Flood-hazard Zone (A Zone) 19
    High-hazard Zone (V Zone) 19
Flood-level rim 19
Floor sink 19
Flow pressure 19
Flue 19
Flush tank 19
Flushometer tank 19
Flushometer valve 19
Foot valve 19
Free circulation 19
French drain 19
Funnel fitting 19

G

Galvanized iron 19
Gate valve 19
Globe valve 19
Grade 19
Grease interceptor 8, 19
Grease trap 20
Ground joint union 20
Groundwork 20
Gutter 20

H

Handhole plug 20
Hangers 20
Head pressure 20
History 1
Horizontal branch drain 20
Horizontal pipe 20
Hose bibb 20
Hot water 20
House drain 20
House sewer 20
House trap 20
Hydrostatic test 20

I

Icemaker box 20
Impervious 20
Indirect drains 58
Indirect waste pipe 20
Individual sewage disposal system 20
Individual vent 20
Individual water supply 20
Inspections 39
Interceptor 20

INDEX

Invert elevation 21
Isometric drawing 21

J

Joint 21
    Expansion 21
    Flexible 21
    Mechanical 21
    Slip 21

K

Knockout plug 21

L

Lateral 21
Lavatory 21
Leader 21
Lead-free pipe and fittings 21
Lead-free solder and flux 21
Liquid petroleum gas (LPG) 21
Listing agency 21
Local vent stack 21
Loop vent 21

M

Main 21
Main vent 21
Maintenance of hot water 6
Manhole 21
Manifold 21
Materials 34
Maximum flow rates 5
Mechanical joint 21
Medical gas system 21
Medical vacuum systems 21
Minimum pipe size 5
Minor repairs 22
Mop sink 22

N

National Pipe Thread (NPT) 22
Nonpotable water 22
Notching and boring of walls 35
Nuisance 22
Number of fixtures 72

O

Occupancy 22
Offset 22
Open air 22
Orifice 22
O-ring 22
Overflow 22

P

Packing 22

Permits 30
Pipe covering 22
Piping material 6
Pitch of drain pipe 7
Plans 34
Plumbing 22
Plumbing appliance 23
Plumbing appurtenance 23
Plumbing fixture 23
Plumbing system 23
Pollution 23
Potable water 23
Pressure gauge 23
Pressure regulator 23
Private 23
Private disposal systems 85
Public or public utilization 23
Public water main 23

Q

Quick-closing valve 23

R

Ready access 23
Reduced pressure principle backflow preventer 23
Referenced standards 4
Registered design professional 23
Relief valve 23
    Pressure-relief 23
    Temperature- and pressure-relief (T&P) 23
    Temperature-relief 23
Relief vent 23
Rim 23
Riser 24
Roof drain 24
Rough-in 24

S

Sanitary 24
Schematics 42
Self-closing faucet 24
Separator 24
Septic tank 24
Sewage 24
Sewage ejectors 24
Sewage pumps and ejectors 8
Sewer 24
    Building sewer 24
    Public sewer 24
    Sanitary sewer 24
    Storm sewer 24
Showers 74
Sill cock 24
Sizing of drainage system 7, 45
Slope 24
Soil pipe 25
Special wastes 58

INDEX

Spill-proof vacuum breaker 25
Stack 25
Stack vent 25
Stack venting 25
Standard pipe weight 25
Standards 32
Sterilizer 25
   Boiling type 25
   Instrument 25
   Pressure (autoclave) 25
   Pressure instrument washer 25
   Utensil 25
   Water 25
Sterilizer vent 25
Storm 25
Storm drain 25
Storm drainage 79
   Below-grade systems 83
   Combined storm and sewer systems 83
   Engineer-designed systems 84
   Leaders 82
   Materials 79
   Overflow design 80
   Roof drain 80
   Scuppers 82
Strapping tape 25
Structure 25
Subsoil drain 25
Sump 25
Sump pump 25
Sump vent 26
Supports 26
Swimming pool 26
System sizing requirements 5

T

Tailpiece 26
Tempered water 26
Third-party certification agency 26
Third-party certified 26
Third-party tested 26
Trap 26
Trap arm 26
Trap seal 26
Type A dwelling unit 26
Type B dwelling unit 26

U

Unstable ground 26

V

Vacuum 26
Vacuum breaker 26

Vent pipe 26
Vent stack 26
Vent system 26
   Acceptable venting methods 8
   Air-admittance valves 9
   Common venting 8
   Vent pipe sizing 9
   Waste stack venting 9
   Wet venting 8
Vent termination 48
Venting 47
Venting methods 51
   Air-admittance valves (Section 917) 53
   Circuit venting (Section 911) 55
   Combination drain and vent system (Section 912) 56
   Common vent (Section 908) 54
   Engineered vent systems (Section 918) 54
   Island fixture venting (Section 913) 51
   Relief vents—stacks for more than 10 branch intervals (Section 914) 52
   Waste stack vent (Philadelphia single stack) (Section 910) 54
   Wet venting (Section 909) 55
Venting systems 8
Vertical pipe 26

W

Wall-hung water closet 26
Waste 26
Waste pipe 27
Water hammer 6
Water heater 27, 74
Water main 27
Water outlet 27
Water services 61
   Sizing 62
   Water hammer 64
Water softener 27
Water-distribution pipe 27
Water-hammer arrester 27
Water-pipe riser 27
Water-service pipe 27
Water-supply system 27
Well 27
   Bored 27
   Drilled 27
   Driven 27
   Dug 27
Wet vent 27
Whirlpool bathtub 27

Y

Yoke vent 27

CODES AND RELATED PRODUCTS

Plumber's Standard Handbook
This complete reference to plumbing codes, installation, troubleshooting and repairs has been prepared by master plumber R. Dodge Woodson and is packed with years of invaluable experience and know-how. It offers a wealth of time-saving charts, tables, drawings and equations. A CD-ROM is included with easy-to-use contracts and forms—all ready to print.
Item No. 113H00

IRC Plumbing Calculators
This helpful three-part set is based on the plumbing provisions of the 2000 *International Residential Code*® (IRC) to assist in sizing. The calculators cover water sizing, drainage and venting. Designed for inspectors, contractors and engineers.
Item No. 111P2K

Pipefitter's Handbook
This helpful tool explains and illustrates the basic problems of pipe bending and fabrication of welded fittings, together with the principles applied in solving those problems. It contains sections on soldering and brazing; types K, L and M copper tube; DWV copper drainage tube; plastic pipe; screwed offset connections; rolling offsets; and more. Numerous illustrations and tables.
Item No. 103H04

Code Check Plumbing
Simplifies the 2000 *International Plumbing Code*® and explains in greater detail the 2000 *Uniform Plumbing Code*™, providing clear answers to the code in effect in your area.
Item No. CDPB2K

International and Uniform Plumbing Codes Handbook
An all-in-one manual written by master plumber and well-known author R. Dodge Woodson that puts the two codes used in the United States into user-friendly language. Users can answer pipe, drainage, vent and trap questions; solve common problems with illustrations of workable solutions; find worked-out examples of nearly every type of plumbing task; quickly locate figures, formulas and charts for water heaters, fixtures and faucets, fuel piping, storm water drainage, and all other calculation needs; and much more! Referencing up to the standards of the new 2000 codes, this is an excellent guide for all plumbers from apprentice to master.
Item No. 112H2K

2000 International Plumbing Code
Developed by the International Code Council (ICC), the *International Plumbing Code*® (IPC) provides consistent requirements that can be used across the country to provide comprehensive regulation of plumbing systems, setting minimum regulations for plumbing facilities in terms of both performance and prescriptive objectives, the IPC provides for the acceptance of new and innovative products, materials and systems. Features 14 chapters and appendices covering all major topics including water sizing, vent sizing, and drain waste. The loose-leaf edition is in a durable seven-hole binder and includes sheet lifters imprinted with rulers and conversion charts. *(131 pages)*
Item No. 112L2K

2000 International Private Sewage Disposal Code
Published as a companion document to the *International Plumbing Code*®, the *International Private Sewage Disposal Code*® (IPSDC) provides flexibility in the development of safe and sanitary individual sewage disposal systems and includes detailed provisions for all aspects of design, installation and inspection of private sewage disposal systems. Published in a soft-cover format that is seven-hole punched for easy storage in an IPC binder. *(74 pages)*
Item No. 012S2K

2000 International Plumbing and Mechanical Codes CD-ROM
Navigate easily through the 2000 IPC and 2000 IMC. Search text, copy images from figures and tables, copy-and-paste code provisions into correspondence or reports, add notes directly to text in color or hide them, and more!
Item No. 103C2K

International Plumbing Code Video Training Series
(Set of 4 Tapes & Workbook - Based on the 1997 IPC)
A new video series, based on the 1997 *International Plumbing Code*® (IPC), includes four videos and a workbook covering:
- Vents and Venting
- Water Heaters
- Methods of Protecting Against Backflow
- Sanitary Drainage

The IPC Video Training Series Workbook is based on chapters 5, 6, 7 and 9 of the 1997 IPC. Each section contains a review of the code requirements followed by questions and answers to strengthen your knowledge. *(46 pages)*
Item No. 297X97

1997 International Plumbing Code Workbook
This workbook is intended to provide practical learning assignments for independent study of the provisions of the 1997 *International Plumbing Code*® (IPC). The independent study format provides a method for the individual to complete the course in an unlimited amount of time. The workbook can also be used for student assignments when integrated as part of a classroom program.
Item No.214W97

Plumber's Standard Handbook
The only plumbing guide you'll ever need, *Plumber's Standard Handbook* provides complete coverage on codes, installation, troubleshooting, and repairs. With a wealth of clear, time-saving charts, tables, drawings, equations, and more.
Item No. 113H00

EXTENDEX – An Extended Index to the IPC
This series of extended indexes has been developed by ICBO to assist users in the transition to the 2000 International Code Series. Each Extendex makes it easy to find what you are looking for in the 2000 IPC, IMC, IFGC or IECC by listing the page and section in the International Code where a subject is covered.
Item No. 112E2K

UPC-IPC Comparison & Cross Reference

This helpful tools provide a comparative analysis between the 1997 *Uniform Plumbing Code*™ (UPC) and the 2000 *International Plumbing Code*® (IPC) with a cross-reference directory. Code sections, tables and figures from the IPC are listed sequentially with an analysis of comparison to the UPC that can be used as a plan checking and field inspection aid. Readers can also locate IPC section numbers for similar provisions found in the UPC and transition them to the IPC.
Item No. 112CCR

2000 IPC Workbook – A Study Companion

This workbook is intended to provide practical learning assignments for independent study of the provisions of the 2000 *International Plumbing Code*. The independent study format provides a means for an individual to complete the course in an unlimited amount of time. The workbook can also be used for student assignments when integrated as part of a classroom program.